iPhone App Development

THE MISSING MANUAL

*The book that
should have been
in the box*®

iPhone App Development

Craig Hockenberry

POGUE PRESS™
O'REILLY®

Beijing · Cambridge · Farnham · Köln · Sebastopol · Taipei · Tokyo

iPhone App Development: The Missing Manual
by Craig Hockenberry

Printed in the United States of America.

Published by O'Reilly Media, Inc., 1005 Gravenstein Highway North, Sebastopol, CA 95472.

O'Reilly Media books may be purchased for educational, business, or sales promotional use. Online editions are also available for most titles: my.safaribooksonline.com. For more information, contact our corporate/institutional sales department: 800-998-9938 or *corporate@oreilly.com*.

May 2010:   First Edition.

This book uses a durable and flexible lay-flat binding.

ISBN: 9780596809775

[M]

Table of Contents

Part Two: Development in Depth

Part Four: Appendix

The Missing Credits

About the Author

Craig Hockenberry has been designing award-winning software for over 30 years. He is currently a principal at the Iconfactory, a company that has been changing the face of the computers since 1996. Their work includes the design and production of icons for Microsoft, Apple, Adobe, and other leading software companies.

Craig almost didn't buy an iPhone. But a mere five minutes at his local Apple store convinced him that he was holding the future in his hands. Shortly thereafter, curiosity led to a hacked iPhone where he could run his own applications. Apple's introduction of the App Store let Craig and his fellow factory workers share that software with the rest of the world.

Since those early days, the Iconfactory (*www.iconfactory.com*) has released many groundbreaking and successful titles for the iPhone, including Twitterrific, Frenzic, Ramp Champ, and Pickin' Time. The company has also contributed design work to many other leading apps.

Craig loves writing and hopes you will learn to share his fascination with BIG LETTERS.

About the Creative Team

Nan Barber (editor) has been involved with the Missing Manual series since its inception. Her job lets her combine her fascination with bright shiny things and lifelong love affair with the written word. Email: *nanbarber@oreilly.com*.

Nellie McKesson (production editor) is a graduate of St. John's College in Santa Fe, New Mexico. She lives in Brockton, Mass., and spends her spare time studying graphic design and making t-shirts (*www.endplasticdesigns.com*). Email: *nellie@oreilly.com*.

Matt Drance (technical reviewer) is the owner of Bookhouse Software, where he builds iPhone apps for himself and for clients. Before founding Bookhouse, Matt spent eight years in Apple Evangelism and DTS, working closely with hundreds of third-party developers to build many of the iPhone apps you use today. He happily lives, works, and races cars in Northern California.

Tina Spargo (technical reviewer), her husband (and professional musician) Ed, their preschooler Max, their two silly Spaniels, Parker (Clumber), and Piper (Sussex), all share time and space in their suburban Boston home. Tina juggles being an at-home mom with promoting and marketing Ed's musical projects and freelancing as a virtual assistant. Tina has over 20 years' experience supporting top-level executives in a variety of industries. Website: *www.tinaspargo.com*.

Jan Jue (copy editor) enjoys freelance copyediting, a good mystery, and the search for the perfect potsticker. Email: *jjuepub@sbcglobal.net*.

Angela Howard (indexer) has been indexing for over 10 years, mostly for computer books, but occasionally for books on other topics such as travel, alternative medicine, and leopard geckos. She lives in California with her husband, daughter, and two cats.

Acknowledgements

First, I must thank my fellow factory workers for allowing me the time to write this book. Dave, Ged, Corey, Talos, Cheryl, Kate, Sean, David, Louie, Anthony, Mindy, and Travis, I feel honored to have you all as colleagues.

I also want to thank Lucas Newman for providing the inspiration with Lights Off and for teaching me how to break out of jail. All the talented people at Apple deserve recognition, too. Without their hard work, none of this would be happening.

The enthusiasm of my good friends, Kim and Mirella Poindexter, got me excited about the project and Jeffrey Zeldman deserves credit for explaining why seven long months of hard labor was a good idea. Along the way, Matt Drance and Tina Spargo kept me honest and taught me new things. The erudite Nan Barber made me look smarter than I really am.

Of course, none of this would have happened without Bill and Mary Jay Hockenberry, who let me be creative even if it meant a trip to the emergency room. Thanks to my brother Kevin and sister-in-law Chris, for the constant reminder that normal people use software. And last, but in no way least, heartfelt gratitude goes to Lauren Mayes for her love, opinions, sense of humor and interpretive dance.

The Missing Manual Series

Missing Manuals are witty, superbly written guides to computer products that don't come with printed manuals (which is just about all of them). Each book features a handcrafted index; cross-references to specific pages (not just chapters); and Rep-Kover, a detached-spine binding that lets the book lie perfectly flat without the assistance of weights or cinder blocks.

Recent and upcoming titles include:

Access 2007: The Missing Manual by Matthew MacDonald

Access 2010: The Missing Manual by Matthew MacDonald

Buying a Home: The Missing Manual by Nancy Conner

CSS: The Missing Manual, Second Edition, by David Sawyer McFarland

Creating a Web Site: The Missing Manual, Second Edition, by Matthew MacDonald

David Pogue's Digital Photography: The Missing Manual by David Pogue

Dreamweaver CS4: The Missing Manual by David Sawyer McFarland

Dreamweaver CS5: The Missing Manual by David Sawyer McFarland

Excel 2007: The Missing Manual by Matthew MacDonald

Excel 2010: The Missing Manual by Matthew MacDonald

Facebook: The Missing Manual, Second Edition by E.A. Vander Veer

FileMaker Pro 10: The Missing Manual by Susan Prosser and Geoff Coffey

FileMaker Pro 11: The Missing Manual by Susan Prosser and Stuart Gripman

Flash CS4: The Missing Manual by Chris Grover with E.A. Vander Veer

Flash CS5: The Missing Manual by Chris Grover

Google Apps: The Missing Manual by Nancy Conner

The Internet: The Missing Manual by David Pogue and J.D. Biersdorfer

iMovie '08 & iDVD: The Missing Manual by David Pogue

iMovie '09 & iDVD: The Missing Manual by David Pogue and Aaron Miller

iPad: The Missing Manual by J.D. Biersdorfer and David Pogue

iPhone: The Missing Manual, Second Edition by David Pogue

iPhone App Development: The Missing Manual by Craig Hockenberry

iPhoto '08: The Missing Manual by David Pogue

iPhoto '09: The Missing Manual by David Pogue and J.D. Biersdorfer

iPod: The Missing Manual, Eighth Edition by J.D. Biersdorfer and David Pogue

JavaScript: The Missing Manual by David Sawyer McFarland

Living Green: The Missing Manual by Nancy Conner

Mac OS X: The Missing Manual, Leopard Edition by David Pogue

Mac OS X Snow Leopard: The Missing Manual by David Pogue

Microsoft Project 2007: The Missing Manual by Bonnie Biafore

Microsoft Project 2010: The Missing Manual by Bonnie Biafore

Netbooks: The Missing Manual by J.D. Biersdorfer

Office 2007: The Missing Manual by Chris Grover, Matthew MacDonald, and E.A. Vander Veer

Office 2010: The Missing Manual by Nancy Connor, Chris Grover, and Matthew MacDonald

Office 2008 for Macintosh: The Missing Manual by Jim Elferdink

Palm Pre: The Missing Manual by Ed Baig

PCs: The Missing Manual by Andy Rathbone

Personal Investing: The Missing Manual by Bonnie Biafore

Photoshop CS4: The Missing Manual by Lesa Snider

Photoshop CS5: The Missing Manual by Lesa Snider

Photoshop Elements 7: The Missing Manual by Barbara Brundage

Photoshop Elements 8 for Mac: The Missing Manual by Barbara Brundage

Photoshop Elements 8 for Windows: The Missing Manual by Barbara Brundage

PowerPoint 2007: The Missing Manual by E.A. Vander Veer

Premiere Elements 8: The Missing Manual by Chris Grover

QuickBase: The Missing Manual by Nancy Conner

QuickBooks 2010: The Missing Manual by Bonnie Biafore

QuickBooks 2011: The Missing Manual by Bonnie Biafore

Quicken 2009: The Missing Manual by Bonnie Biafore

Switching to the Mac: The Missing Manual, Leopard Edition by David Pogue

Switching to the Mac: The Missing Manual, Snow Leopard Edition by David Pogue

Wikipedia: The Missing Manual by John Broughton

Windows XP Home Edition: The Missing Manual, Second Edition by David Pogue

Windows XP Pro: The Missing Manual, Second Edition by David Pogue, Craig Zacker, and Linda Zacker

Windows Vista: The Missing Manual by David Pogue

Windows 7: The Missing Manual by David Pogue

Word 2007: The Missing Manual by Chris Grover

Your Body: The Missing Manual by Matthew MacDonald

Your Brain: The Missing Manual by Matthew MacDonald

Your Money: The Missing Manual by J.D. Roth

Introduction

These days, there's no shortage of books about how to develop an iPhone app. But to make your product a success, you need to do much more than just create great software.

iPhone App Development: The Missing Manual takes you through the entire development process. You'll learn how to write the code for a successful title on the App Store, but just as importantly, you'll acquire the skills to design, test, and market that product.

Unlike other books that take a dry, mechanical approach to the topic, this book tells the story of a real product's development from start to finish. You'll follow along as an actual iPhone developer recounts the tale using the popular Safety Light application as a protagonist.

Creating a great iPhone app is often a group effort. Whether you're a developer, designer, marketer, or project manager, you'll find topics that get you up to speed on this new and exciting platform. And when there's more to learn, expert advice will point you in the right direction to fill in the details.

The App Store

Since the iTunes App Store's launch in July 2008, over 100,000 iPhone applications have been submitted to the store. Customers have downloaded over 3 billion applications that were created by developers just like you. The success of this endeavor has exceeded everyone's wildest expectations.

Before the App Store was launched, iPhone app development was limited to the talented engineers at Apple's headquarters in Cupertino, California. But in just a couple of years, thousands of developers worldwide have discovered how easy and fun it is to write software for the iPhone. And by keeping 70 cents of every dollar spent on their app in iTunes, some developers have found these apps to be very profitable.

These early adopters also learned something the hard way: This new and innovative computing device that you carry around in your pocket comes with a different set of rules. A multitouch display with ubiquitous networking in a small form presents many challenges.

The difficulties are not limited to technology, either. How you design, build, and distribute your apps requires a new way of thinking. Many developers have struggled with their initiation into a consumer mass market.

As you walk through the iPhone app development process from start to finish, you'll learn from those who preceded you. You'll avoid the pitfalls of some, while learning from the success of others. The goal, of course, is to help you make the best application possible.

Figure 1-1:
What better way to learn iPhone app development than to watch an experienced developer build a product with step-by-step instruction? In this book, you'll see the Safety Light (A) app come to life and go on sale in the iTunes App Store.

About This Book

Despite the many improvements in software over the years, one feature has grown consistently worse—documentation. With the purchase of most software programs these days, you don't get a single page of printed instructions. To learn about the hundreds of features in a program, you're expected to use online electronic help.

But even if you're comfortable reading a help screen in one window as you try to work in another, something is still missing. At times, the terse electronic help screens assume you already understand the discussion at hand, and hurriedly skip over important topics that require an in-depth presentation. In addition, you don't always get an objective evaluation of the program's features. (Your fellow engineers often add technically sophisticated features to a program because they *can*, not because you need them.) You shouldn't have to waste your time learning features that don't help you get your work done.

In this book's pages, you'll find step-by-step instructions for developing iPhone applications. In addition, you'll find that "big picture" topics such as design, sales, and marketing are covered. The goal is to make you an effective and successful developer, not just to teach you how to write the code.

Note: This book periodically recommends *other* books, covering topics that are too specialized or tangential for a manual about iPhone development. Careful readers may notice that not every one of these titles is published by Missing Manual–parent, O'Reilly Media. While we're happy to mention other Missing Manuals and books in the O'Reilly family, if a great book out there doesn't happen to be published by O'Reilly, we'll still let you know about it.

---

iPhone App Development: The Missing Manual is designed to accommodate readers at different technical levels. The primary discussions are written for computer users with some programming knowledge. But if you're a first-timer, special sidebar articles called "Up to Speed" provide the introductory information you need to understand the topic at hand. If you're an advanced user, on the other hand, keep your eye out for similar shaded boxes called "Power Users' Clinic." They offer more technical tips, tricks, and shortcuts for the experienced developer.

About the Outline

iPhone App Development: The Missing Manual is divided into four parts, most containing several chapters:

- **Part 1: Getting Started with Cocoa Touch.** In the first four chapters, you'll build your first iPhone App and get acquainted with your basic tools: Cocoa Touch, Interface Builder, Xcode, and the Objective-C programming language. You'll also start thinking about how to use these tools to design a new application.

- **Part 2: Development in Depth.** In the next three chapters, you'll learn how to set up your iPhone development environment, including getting your app onto a phone for the first time. You'll also take a guided tour through the code of the finished app and learn how to test the final product.

- **Part 3: The Business End.** The final two chapters explore the business of being an iPhone developer. You'll learn how to get your app onto iTunes, promote it through various marketing channels, and how to keep track of your sales. A survey of the iPhone app market will help you understand where your app fits in.

- **Part 4: Appendix.** The appendix introduces you to the vast array of resources for learning more about all of the topics covered in this book.

At the Missing Manual website, you'll find free, downloadable bonus material. In addition to the project and source code for the Safety Light iPhone application, you'll find a promotional website template that you can use for your products.

The Very Basics

This book contains very little jargon or nerd terminology. You will, however, encounter a few terms and concepts that you'll come across frequently in your computing life:

- **Clicking.** This book gives you three kinds of instructions that require you to use your computer's mouse or trackpad. To *click* means to point the arrow cursor at something on the screen and then—without moving the cursor at all—to press and release the left clicker button on the mouse (or laptop trackpad). To *double-click,* of course, means to click twice in rapid succession, again without moving the cursor at all. And to *drag* means to move the cursor while pressing the left button continuously.

- **Keyboard shortcuts.** Every time you take your hand off the keyboard to move the mouse, you lose time and potentially disrupt your creative flow. That's why many experienced developers use keystroke combinations instead of menu commands wherever possible. ⌘-B, for example, is a keyboard shortcut to build your application in Xcode.

  When you see a shortcut like ⌘-S (which saves changes to the current document), it's telling you to hold down the ⌘ key, and, while it's down, to press the letter S key, and then release both keys.

- **Choice is good.** Xcode and Interface Builder frequently give you several ways to trigger a particular command—by choosing a menu command, *or* by clicking a toolbar button, *or* by pressing a key combination, for example. Some people prefer the speed of keyboard shortcuts; others like the satisfaction of a visual command array available in menus or toolbars. This book lists all of the alternatives, but by no means are you expected to memorize all of them.

About→These→Arrows

Throughout this book, and throughout the Missing Manual series, you'll find sentences like this one: "Open the Hard Drive→Developer→Applications folder." That's shorthand for a much longer instruction that directs you to open two nested folders in sequence, like this: "On your hard drive, you'll find a folder called Developer. Open that. Inside the Developer window is a folder called Applications; double-click it to open it."

Similarly, this kind of arrow shorthand helps to simplify choosing commands in menus, as shown in Figure I-2.

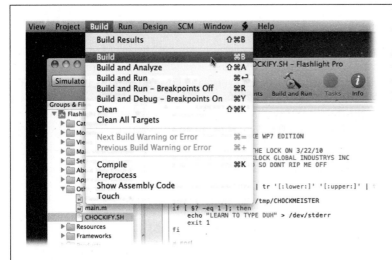

Figure 1-2:
When you read in a Missing Manual, "Choose Build→Build," that means: "Click the Build menu to open it. Then click Build in that menu."

Living Examples

This book is designed to get your work onto an iPhone faster and more professionally; it's only natural, then, that half the value of this book also lies on the iPhone.

As you read the chapters, you'll encounter a number of *living examples*—step-by-step tutorials that you can build yourself, using raw materials (like graphics and source code) that you can download from the Missing CD (*www.missingmanuals. com/cds*). You might not gain much by simply reading these step-by-step lessons while relaxing in your hammock. But if you take the time to work through them at the computer, you'll discover that these tutorials give you an unprecedented insight into the way professional iPhone developers build apps.

iPhone development is also a rapidly evolving topic. To keep up with the latest news, check out this book's website at *http://appdevmanual.com*.

About MissingManuals.com

At *www.missingmanuals.com*, you'll find articles, tips, and updates to *iPhone App Development: The Missing Manual*. In fact, we invite and encourage you to submit such corrections and updates yourself. To keep the book as up to date and accurate as possible, each time we print more copies of this book, we'll make any confirmed corrections you've suggested. We'll also note such changes on the website, so that you can mark important corrections into your own copy of the book, if you like. (Go to *www.missingmanuals.com/feedback*, choose the book's name from the pop-up menu, and then click Go to see the changes.)

Also on our Feedback page, you can get expert answers to questions that come to you while reading this book, write a book review, and find groups for folks who share your interest in iPhone application development.

We'd love to hear your suggestions for new books in the Missing Manual line. There's a place for that on missingmanuals.com, too. And while you're online, you can also register this book at *www.oreilly.com* (you can jump directly to the registration page by going here: *http://tinyurl.com/yo82k3*). Registering means we can send you updates about this book, and you'll be eligible for special offers like discounts on future editions of *iPhone App Development: The Missing Manual*.

Safari® Books Online

 Safari® Books Online is an on-demand digital library that lets you easily search over 7,500 technology and creative reference books and videos to find the answers you need quickly.

With a subscription, you can read any page and watch any video from our library online. Read books on your cellphone and mobile devices. Access new titles before they're available for print, get exclusive access to manuscripts in development, and post feedback for the authors. Copy and paste code samples, organize your favorites, download chapters, bookmark key sections, create notes, print pages, and benefit from tons of other time-saving features.

O'Reilly Media has uploaded this book to the Safari Books Online service. To have full digital access to this book and others on similar topics from O'Reilly and other publishers, sign up for free at *http://my.safaribooksonline.com*.

Part One: Getting Started with Cocoa Touch

I

Building Your First iPhone App

You have an idea that will lead to fame and fortune on the iTunes App Store. You decide to write an iPhone app. The first and most important task is for you to become comfortable with the tools used to build your products. A Chinese proverb says, "the journey is the reward," and this chapter is all about the journey. In the upcoming pages, you'll experience the entire application development process, start to finish. You'll learn how to set up the software you need, and try your hand at building an app.

But what app? If you do a quick search of the App Store, you'll find no shortage of flashlights. For many aspiring developers, this simple application is a rite of passage, so now's your chance to join this illustrious crowd. Once you see how easy it is to create your own app, you'll wonder why people pay 99¢ for them on iTunes!

Getting the Tools

You can't build anything, including an iPhone app, without tools. Luckily, you can find everything you need on your Mac, or download it for free. Specifically, you need to download and install Xcode development software and the iPhone Software Development Kit (SDK) on your Mac. (And if you don't have a Mac, see the box on the next page.)

Both the Mac and iPhone benefit from a rich set of technologies that have stood the test of time. The iPhone SDK is built upon the infrastructure created by NeXT in the 1980s. This company, founded by Steve Jobs, created a revolutionary object-oriented operating system called NeXTSTEP. This influential system has evolved into the OS X operating system in use today. As you learn more about the iPhone, you'll see that it has much in common with the Mac.

Get a Mac

If you're going to create iPhone applications, you're going to do it on a Macintosh. Apple's development tools don't run on Windows or any other operating system. Just as you can't run Microsoft Visual Studio on a Mac, you need a Mac to run the tools used to build your iPhone app. They rely on features of the underlying system software.

If you don't have a Mac, here are some hints to help you make the right purchase:

- **Buy a used machine.** If you're on a shoestring budget, check out eBay or craigslist. Someone else's old hardware will be perfectly fine for iPhone development. The apps you're going to create are small and don't need a lot of processor power to build and test. The only caveat when buying older hardware is to make sure the Mac has an Intel processor. The development tools don't work with older PowerPC processors.

- **Add a Mac mini.** Buying a new Mac mini is a great option if you already have a display, keyboard, and other peripherals. You can save quite a bit of money by just buying a new CPU and repurposing the devices you already own. If you're a software developer, you probably have this stuff already sitting around in a closet. And if you're developing for multiple platforms,

it's handy to put the Mac mini behind a KVM switch so you can quickly shift between machines.

- **Go ahead and splurge.** Apple makes some very sexy hardware. In particular, the new laptops are hard to resist. If you're looking for excuses to justify the purchase, here's some help:

Macs now use an Intel processor, which means you can run Windows or any other x86-based operating system on your new machine. You can boot into any operating system using Apple's free Boot Camp utility. Or you may find it easier to install third-party software like VMware Fusion and to run other operating systems on a virtual machine within Mac OS X.

Virtual machines are particularly handy when you need to see how your iPhone product website appears in Internet Explorer. Just launch the virtual machine, open the browser in Windows, and load the test URL.

Finally, think of all the money you're saving on development tools. If you're used to spending thousands of dollars on Visual Studio and MSDN, it will come as a pleasant surprise to know that all of Apple's developer tools are free. Spend your dollars on the hardware instead of the tools and you'll come out ahead.

Installing Xcode

Once you and your Mac are ready to go, it's time to load your hard drive with lots of new software. Apple supplies the Xcode development tools free of charge, but doesn't install them on every Mac, since most consumers will never use them.

Luckily, you can find the Xcode tools right on your Snow Leopard installation disk. To run Xcode, Apple recommends you have an Intel-based Mac running Leopard or Snow Leopard. The following steps explain how to get the software onto your hard drive where you can use it:

Note: You can install the iPhone SDK and other development tools on Leopard, but the Snow Leopard tools reflect significant improvements over the previous version. Working in the newest version of Mac OS X assures you the latest and greatest features.

---

1. **Pop the installation DVD into your Mac and double-click its icon. In the Optional Installs folder, double-click the Xcode.mpkg file.**

   When you double-click that file, the Xcode installation process begins.

2. **On the introductory screen, click Continue. When the license agreement screen appears, click Continue and then click Agree.**

   The license agreement is the same legalese you agree to whenever you install software. Read it if you're into such things. When you're done, the next screen lets you choose what you want to install, as shown in Figure 1-1.

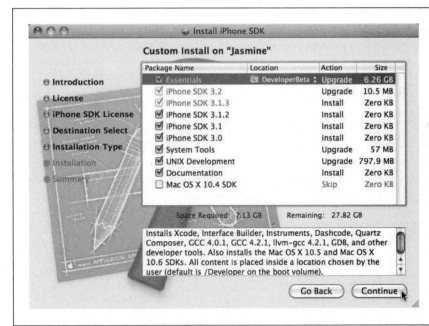

Figure 1-1:
The Xcode installer starts out with all checkboxes except Mac OS X 10.4 selected for you. Leave them that way. You can click each package name to see what's being installed: Besides the integrated development environment (IDE) that you'd expect, you'll also find tools for monitoring performance and plenty of documentation.

3. **On the Custom Install screen, make sure all checkboxes except Mac OS X 10.4 are turned on, and then click Continue.**

   The installer copies all of the Xcode from the DVD onto your hard drive. This process takes a few minutes.

4. **The final screen prompts you to select an install location. Make sure that it's on the same disk where all your other applications are stored. Click Install to start the process.**

   Depending on your Mac's speed, this process can take a few hours. Get away from the computer and get some fresh air for once.

   After everything is safely on your hard drive, you see this message: "The installation was successful."

5. **Click Close to quit.**

6. **You can safely eject the DVD at this point.**

After the installation is complete, go to the Hard Drive→Developer→Applications folder on your hard drive, and check out your new tools. This folder contains the applications and utilities you use to develop both Mac and iPhone applications: The ones you'll use the most are Xcode and Interface Builder. The parent Developer folder also has all of the accompanying developer frameworks, libraries, and documentation.

The Xcode installation doesn't include one thing—the iPhone SDK that's required to develop apps for your phone. For that, go on to the next section.

Tip: Now that you have your tools, maintain them. Apple regularly updates Xcode, so the version on your Snow Leopard DVD will eventually become outdated. When major changes occur, Apple will send an email reminding you to upgrade by visiting the iPhone Dev Center, as described in the next section.

Getting the iPhone SDK

You have to join the iPhone Developer Program before Apple lets you get your hands on the iPhone SDK. Your free membership gives you access to the tools, documentation, and developer forums via the iPhone Dev Center (Figure 1-2).

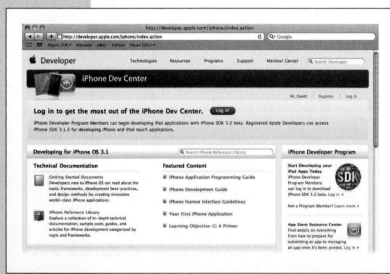

Figure 1-2:
The iPhone Dev Center is your first and best resource as an iPhone developer. You'll use this site to download and update your iPhone SDK, find sample code and documentation, connect with other iPhone developers, and to prepare your product for sale on iTunes.

1. To sign up for an ADC membership, point your web browser to *http://developer.apple.com/iphone/*. Click the Register link in the upper-right corner.

   You access the iPhone Dev Center using an Apple ID. If you have an iTunes account or have made a purchase from the Apple Store, you already have one set up. Go ahead and use it when you create your developer account and skip to step 4.

Note: If you've been using your Apple ID for personal stuff like iTunes and a MobileMe family photo gallery, you may want to create a new Apple ID for your developer account. Having a separate Apple ID used solely for business purposes can help you avoid accounting and reporting issues. Please see Chapter 8 to see how your developer account and iTunes Connect affect your business.

---

2. If you're setting up a new Apple ID, type your name, contact information, and security questions for password retrieval.

3. Turn on the checkbox to accept the licensing agreement and click Continue.

   In a few minutes, Apple will send you an email to verify the account.

4. Click the Email Verification link, and enter the code contained in the message to complete the account setup.

Once you set up your account and log in, you see a lot of new content available from the iPhone Dev Center. You have access to great resources like the Getting Started Videos, Coding How-To's, and Sample Code. Right now, turn your attention to the download for the iPhone SDK.

1. Click the Downloads link, and you see a selection of links at the bottom of the page, as shown in Figure 1-3.

   As new versions of the iPhone SDK are released, these links will be updated. Pick the most recent release that matches your version of Mac OS X. At the time of this writing, it's "iPhone SDK 3.1.3 with Xcode 3.2.1".

   The iPhone SDK is a large download: Its size can range from several hundred megabytes to over 2 GB. Be patient as it downloads from your web browser, is verified, and mounted: It's going to take a while.

   Once it's finished, you have a .dmg disk image in your Downloads folder and a new iPhone SDK disk on your desktop, as shown in Figure 1-4.

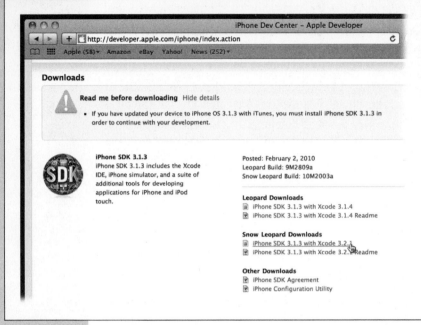

Figure 1-3:
You find the links to download the iPhone SDK toward the bottom of the iPhone Dev Center page. The links in this picture are for version 3.1.3, but these will change as Apple updates the SDK. You can click the Read Me links to see what's new in the release.

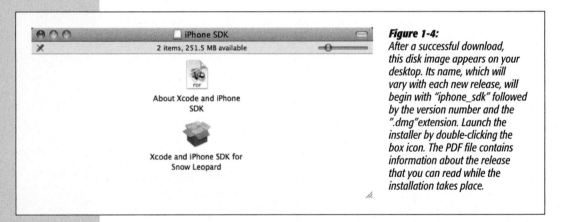

Figure 1-4:
After a successful download, this disk image appears on your desktop. Its name, which will vary with each new release, will begin with "iphone_sdk" followed by the version number and the ".dmg" extension. Launch the installer by double-clicking the box icon. The PDF file contains information about the release that you can read while the installation takes place.

Once you have the iPhone SDK disk image, you can begin the installation:

1. Double-click the "iPhone SDK" file to start the installation process. It's the brown and gold box icon.

2. Click Continue on the welcome and license agreement screens.

3. Click Agree to accept the license.

4. On the Install screen, click Continue to install the standard packages, and then click Install to start the installation process.

5. If required, enter your password so system files can be modified.

6. It's also a good idea to quit iTunes at this point to avoid a dialog box that pauses the install.

   Depending on your Mac's speed and the size of the download, the installation process can take anywhere from half an hour to several hours. When the installation is finished, you'll see a green checkmark and can click Close to finish. At this point you can eject the iPhone SDK disk, but keep the .dmg file around as a backup.

Note: As with Xcode, Apple updates the iPhone SDK regularly. You'll need to return to this iPhone Dev Center periodically to install the latest version of the SDK. Apple typically releases a new version in conjunction with a new iPhone firmware release.

What Lies Ahead for the SDK?

The iPhone SDK is constantly evolving as Apple fixes bugs and adds new features. You'll want to update your development environment to keep up with the latest changes. Apple updates the iPhone SDK in two different ways. The first, and simplest, is a maintenance release. These releases just fix bugs in the firmware and don't introduce any new features. In most cases, you won't need to make any changes to your application.

Apple provides maintenance releases of the SDK to developers on the same day that it makes the firmware available to customers. These releases have a three-part version number like 2.2.1 and 3.1.3. As soon as you install the new firmware on your device, you need to update the iPhone SDK so you can install and debug your applications from Xcode. If you don't, you'll see warnings that the tools don't support the device's firmware version.

When Apple makes more substantial firmware changes that will affect developer software, either by adding new features or changing existing ones, it posts a beta version of the iPhone SDK on the iPhone Dev Center. Only developers who have paid to join the iPhone Developer Program have access to these advance releases. (Page 30 explains how to enroll in this developer program.) These betas are for major releases, such as 3.0 or 4.0, or revisions like 3.1. Apple typically starts the beta release cycle three or four months prior to a general public release. Once the cycle starts, it puts out a new SDK (called Beta 1, Beta 2, and so on) every couple of weeks. These beta releases usually also include a new version of Xcode with improvements and support for the new iPhone OS, along with new firmware.

With early access to the new SDK, you can build and run your application with the new iPhone firmware. If you've been careful to use only documented features and APIs, you shouldn't have many issues to deal with: Apple's engineers are very good at maintaining compatibility with published interfaces. You may see deprecation warnings as you compile, but those are usually simple to fix. It's more likely that you'll spend the beta test period learning about new features and testing them out in your application.

There are a couple caveats to keep in mind when installing a beta version of the iPhone SDK. First, you can't use the beta tools to submit an application to the App Store. Luckily, you can install multiple versions of Xcode on your hard drive. To install the tools in a separate location, follow these steps:

1. **Quit the iPhone Simulator if it's running.**

   If you skip this step, the installation process will hang indefinitely, and you'll need to quit the Installer and start over.

2. **Double-click the iPhone SDK icon in the disk image to start the installation process. Agree to the licenses and choose a destination hard drive.**

   You see a list of packages to install. In the second column, Developer is set as the Location. You need to change the location for the beta release.

3. **Click Developer and then select Other from the pop-up menu (see Figure 1-5).**

   A dialog box opens for you to select a folder.

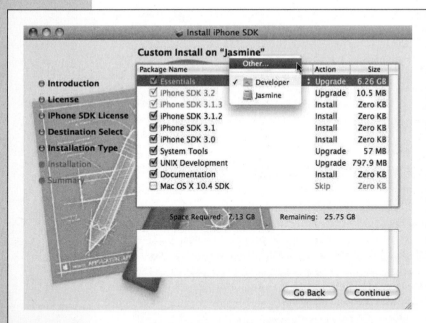

Figure 1-5:
You can choose a custom installation location for the iPhone SDK. Since you can't use beta releases of the iPhone SDK to build your application for the App Store, you'll need to keep two versions of the tools on your hard drive. During the install process, click the Developer folder icon and select Other to choose the location for the beta version.

4. Navigate to the root of your hard drive by selecting its name from the list of DEVICES. Then click the New Folder button and type *DeveloperBeta*. Click Create to create the folder.

5. Select Choose to use the *DeveloperBeta* folder for the installation.

   After you return to the main installation window, you'll see DeveloperBeta as the Location.

6. To use the beta, launch Xcode and other tools from the new Hard Drive→ DeveloperBeta→Applications folder.

Now for the second caveat: the beta release is Apple Confidential Information and is covered by a Non-Disclosure Agreement (NDA). These big legal words mean that you can't talk about it in public. You can discuss the new SDK only on the Apple Developer Forums (*http://devforums.apple.com*). You can connect with other developers who are doing the same thing you are: learning about a new release by asking questions and sharing discoveries. Apple engineers also contribute to the discussion.

The NDA also means that you won't find any books or other media to help you understand the changes. The only information about the beta release comes from Apple itself and is posted on the iPhone Dev Center. Typically there's a "What's New" document, release notes, and a list of API differences. Read each of these documents fully: it's a great way to pass the time when you're waiting for several gigabytes of SDK to download!

Another source of information is Apple's annual developer conference, WWDC. Beta releases often coincide with this weeklong conference so everyone can discuss new features in detail. The conference takes place during the summer in San Francisco: it's a great opportunity to meet your fellow developers and learn lots of new things.

Exploring Your New Tools

Your Mac is now set up to create iPhone applications, so you're ready to start making your first one. The best part is that you're not going to write any code. How can you develop without writing code? It's possible with the timesaving power of Xcode templates and Interface Builder.

If you're an experienced developer, this way of working can present a challenge. If you're used to working in Visual Studio, Eclipse, or some other environment, your first encounter with Xcode can be a bit daunting. Besides working on a new operating system, you're also going to be dealing with new project layouts, keyboard shortcuts, and preferences. Don't worry, all of the tools you're used to having are still there, it's just a matter of time before you become comfortable using Xcode's version of them.

In this section, you'll go through all of the phases of creating an iPhone app, from creating a project file with Xcode to running it in the iPhone Simulator. You'll also take a peek at the Interface Builder application that lets you modify the user interface.

Every Flashlight Needs a Parts List

The first phase of creating an iPhone app is setting up a Project file. This file keeps track of the information Xcode uses to build your application. It's where you manage your source code, user interfaces, frameworks, and libraries. Think of it as a parts list for your application.

1. **In your Hard Drive→Developer→Applications folder, double-click the Xcode icon to start the application. (It's at the bottom of the list.)**

   The tricky part is that Xcode isn't in your normal Applications folder. The installer puts it in the *Developer*→Applications folder. To make it easier to return to Xcode later, store its icon in your Dock.

2. **In the Dock, Control-click the icon and choose Options→Keep in Dock.**

   From then on, you can launch Xcode by simply clicking the Dock icon.

   Once Xcode is running, you'll see its Welcome window, as shown in Figure 1-6.

Tip: If you close the Welcome window by accident, you can reopen it by choosing Help→Welcome to Xcode.

Figure 1-6:
The Xcode launch window. As you create new projects with Xcode, you see them listed on the right. Click the "Create a new Xcode project" button to start your first iPhone application. The "Getting started with Xcode" button opens the documentation viewer and displays a helpful overview of Xcode. The last button is a convenient link to the Dev Centers for the Mac and iPhone.

3. **Click the big "Create a new Xcode project" button.**

   The New Project window opens (Figure 1-7), showing you a choice of template categories. In Xcode, a *template* is a predefined set of source code files, libraries, frameworks, and user interface elements that you use to create different styles of applications.

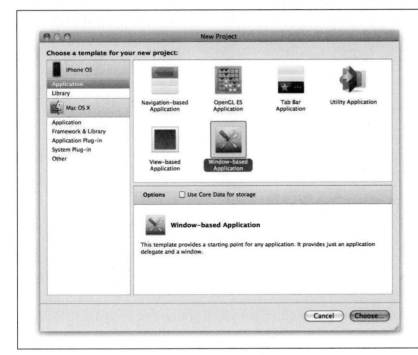

Figure 1-7:
The Xcode New Project window lists all of the templates you can use to get a quick start. When you're starting out with a new application, select the template that best describes the style of user interface you want. When selected, each template displays a short description. Some templates even include options, like the one shown here to "Use Core Data for database storage".

4. Since you're creating an iPhone application, under the iPhone OS group in the upper-left corner, choose Application and take a look at the available templates.

Your choices come in the following categories:

- **Navigation-based Application.** These applications have a "drill-down" style interface, like the iPhone Mail application.

- **OpenGL ES Application.** Games that draw objects in a 3-D space use this template.

- **Tab Bar Application.** This style of application uses a tab bar at the bottom of the screen to switch views. Apple's iPod application is a great example of this user interface style.

- **Utility Application.** These applications generally present a simple interface, with a front view containing information and a back view for configuring the information. The built-in Weather app uses this metaphor.

- **View- and Window-based Applications.** Turn to these templates when your application combines elements of the previous four styles. Think of them as bare-bones templates that you can customize to your own needs.

For your Flashlight app, you're going to use the Window-based Application template. Since the application only uses a single window, this basic template is all you need. A nice side effect of using this customizable template is that it creates fewer files for the project. In effect, you have a shorter, simpler parts list.

5. **Click Window-based Application and then click Choose. Leave the Use Core Data checkbox unchecked since a flashlight doesn't need a database.**

   A save file dialog box appears so you can indicate a name and location for your project's folder (Figure 1-8).

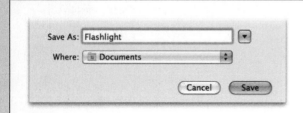

Figure 1-8:
For the Flashlight project, tell Xcode to create a folder named Flashlight inside your Documents folder.

6. **Type *Flashlight* for the project name, and choose the Documents folder from the bottom pop-up menu.**

   As shown in Figure 1-9, the folder you've just created in the Finder contains everything you need to build your application, including the main project file Flashlight.xcodeproj.

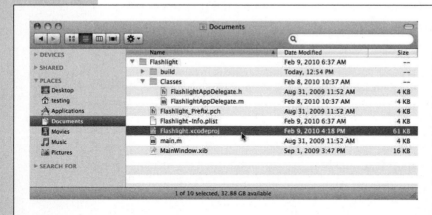

Figure 1-9:
When you create a project with Xcode, it creates a folder of files used to build your application. The most important one is the .xcodeproj file—you can double-click this file to open the project in Xcode. Also, since you're creating a window-based app, Xcode starts you out with a file called MainWindow. xib. You'll learn more about the other files and folders in upcoming chapters.

Xcode creates this project folder behind the scenes. You may never in your Xcode career interact directly with files or folders. Instead, you can rely on Xcode to manage everything for you. But you still need to know where the folder is so you can back up your work.

Tip: As you get more advanced with Xcode, you may want to put your projects in folders of their own within your Home folder. Many developers create a Projects folder that contains nothing but their Xcode folders. Just as the Pictures, Movies, and Music folders make it easier to manage your media, a Projects folder makes it easier to manage your software.

After Xcode finishes creating the new project, it displays the files in a project window, as shown in Figure 1-10.

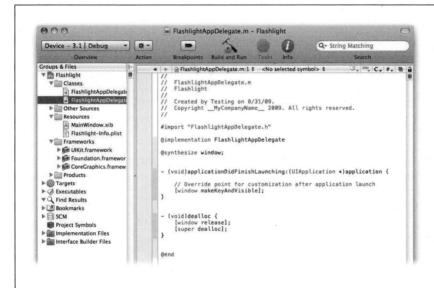

Figure 1-10:
The Xcode project window. On the left are the project's groups and files, and on the right is the source code editor. Although the yellow group icons look like folders, they're not the same as the blue ones you see in the Finder. You can rename the Classes group and not affect the folders on disk. Likewise, you won't find a Resources folder in the Finder, but the group is a great way to organize files that aren't source code.

The Groups & Files panel on the left side of the project window lists the individual files that make up your application. To go back to the parts list metaphor, each group contains parts of a similar type. Here are a few of the most important groups:

- **Classes.** The files in this group contain your project's actual source code.

- **Resources.** User interface files, graphics, and application configuration files all fall under the Resources group.

- **Frameworks.** These files contain tools that the iPhone SDK uses.

The editor part of the window (the big white area in the lower right) shows the code that runs when the application finishes launching. In this simple example, the code makes a window object visible and able to respond to taps.

You'll learn more about these important groups and the source code in the next chapter. Remember: Your goal in this chapter is to build an app without writing any code!

Some Assembly Required

You've gathered all the parts, and now it's time to assemble them. Unlike toys on Christmas Eve, Xcode projects are easy to put together. Thanks to the template, all you have to do is initiate the Build command, and Xcode takes all of the source code, resources, and frameworks and combines them into an executable file that can run on the iPhone.

Make sure that the pop-up menu in the upper-left corner of your project window is set to Simulator and Debug (the exact wording on the menu will change depending on which version of the iPhone SDK you're using). Then choose Build→Build, as shown in Figure 1-11.

After a short wait, you should see "Build succeeded" in the status bar, at lower left.

Note: Xcode's status bar, which runs along the bottom of the window, is an important information source. As you perform various tasks, this area keeps you posted on their progress. It's the first place you should look when you're wondering what's going on.

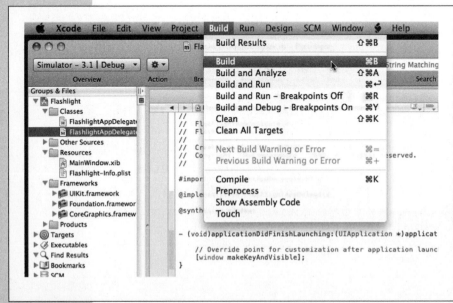

Figure 1-11:
Building your iPhone application. Note that the Overview menu is set to the Simulator and Debug. The keyboard combination ⌘-B is a handy shortcut to build the project after you've made a change to the source code.

Once you've built your app, you can run it on your iPhone. Or better yet, run it on your Mac. That's right. If you're like most developers, you'll run your iPhone apps on the Mac about 90 percent of the time. Apps launch faster on the Mac than on the iPhone, and they're easier to debug on the Mac when problems occur.

Note: Don't discount that last 10 percent. Running your app on an actual device is extremely important: You'll learn why when design issues are discussed in Chapter 4.

Taking It for a Run on Your Mac

When you develop an iPhone app, you'll probably run it on your Mac to test and debug it before it ever gets near an iPhone.

So how do you get an iPhone onto your Mac? It's easy: Make sure Simulator is selected in the Overview menu, and choose Run→Run. Since you have Simulator selected, Xcode uses a simulation of the iPhone to run your app.

Keep an eye down on the status bar. You'll see "Installing Flashlight in Simulator" displayed, and eventually, "Flashlight launched". Soon after, a giant iPhone appears on your desktop, and it's running your Flashlight application (Figure 1-12). Congratulations!

Figure 1-12:
In the iPhone Simulator, the image on the left shows the application running, and the one on the right shows the application's icon on the home screen. It won't fit in your pocket, but the simulator acts just like a real, live iPhone.

Tip: You can do everything you've done in this section and the previous one with a single keystroke. Pressing ⌘-Enter builds and runs your application (in the simulator) in one step. When you get into the thick of iPhone app development, you'll be pressing these keys in your sleep.

So what does this giant iPhone do? It's called the iPhone Simulator, and it behaves like the device in your pocket, except:

- It's hundreds of times faster.
- It has as much memory as your Mac.
- The network is much more reliable.
- It has a larger display.
- It doesn't sync with iTunes over a USB cable.
- Touching the simulator screen has no effect.

In reality, this big phone sitting on your desktop has very different hardware specifications than the one you're used to. But once you get used to keeping your fingers off the screen, you'll end up loving the simulator. It makes your life as a developer so much easier because it does one important thing: it lets you test your code without having an iPhone plugged into your Mac. The box on page 25 shows you how to get the most out of this important tool.

When you click the Home button, you see a screen with the applications installed by Xcode. Since you only have one at this point, all you'll see is the Flashlight app. Drag your mouse to swipe between the pages of applications. Many of the applications you're used to seeing on your iPhone have gone missing, but you'll still be able to use Photos, Contacts, Settings, and Safari while testing. Clicking an app's icon launches it in the Simulator just as it would on a real device.

Tip: The Safari icon in the simulator's tray is very helpful for testing how websites will look on the iPhone. When you start promoting your application, you'll want to use the simulator to check your product pages on a mobile device.

Simulating Reality

The iPhone Simulator acts very much like the device in your pocket. Sometimes, however, it's not obvious how to make the virtual device behave like the physical one.

When you hold down the Option key, two dots appear. These dots show the position of multitouch events when you click the mouse button. Use this feature to simulate the pinching and spreading gestures that zoom the iPhone screen.

You should also check out the Simulator's Hardware menu. Two commands on this menu let you rotate the device left and right—very handy if your app detects device orientation. Choose Hardware→Rotate Left or Rotate Right.

If your application uses shake gestures, there's a menu item to simulate one. Shaking is a standard part of the iPhone copy and paste mechanism (it's used for undo).

The Hardware menu also lets you simulate the status bar that the iPhone displays when you're on a phone call (choose Hardware→Toggle In-Call Status Bar). That way, you can verify that your application resizes its user interface (UI) correctly when the display window is 20 pixels smaller.

(The iPhone displays the same "double height" status bar when using Internet tethering.)

As you get more advanced in iPhone development, you'll have occasion to use the Hardware→Lock command. You can make your app detect when the iPhone sleeps and wakes up, and Lock is the way to simulate that.

Another advanced simulator feature is memory warnings. The iPhone has a limited amount of memory available to applications. Hardware→Simulate Memory Warning lets you test how your application behaves when the phone runs out of memory. Many developers run into problems when they go from an unlimited amount of memory on their Mac to 128 MB on a device. Simulate Memory Warning helps you avoid that fate.

It's also important to realize that the simulator is sharing many of the resources on your Mac. Since both are built on top of OS X technologies, things like the network are common to both platforms: take advantage of the similarity. For example, if you want to see how your app behaves when it loses the cell network, just go into your Mac's System Preferences, and turn off the network interface.

Revision Decision

So now that you have a running application, you notice that the white light makes it look like every other flashlight app in iTunes. You need a better color for the light—a way to stand out in the crowd. How are you going to do that without breaking the "no code" rule? The answer is simple: Interface Builder.

Xcode works hand-in-hand with Interface Builder to create your app's user interface (UI). You create objects like windows and buttons with this tool and then drop them into your code. When you created the project from a template, Xcode created a file containing these objects automatically.

To get an idea of how easy this method is, open your Flashlight app's UI. From the main project window, open the Resources group by clicking the disclosure triangle, and double-click the MainWindow.xib file. Interface Builder opens (you'll see it bouncing in the Dock). Once it's open, you're ready to start working on your UI (Figure 1-13).

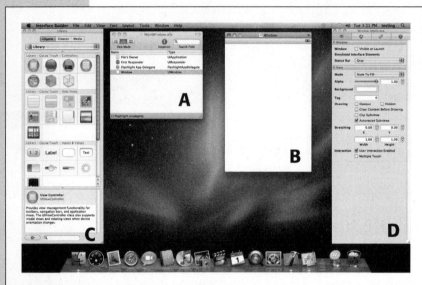

Figure 1-13:
Interface Builder and its many windows take up much of your display. In the middle are the .xib document (A) and the window displayed in the iPhone app (B). To the left is the library of user interface elements where buttons, windows, and other user interface components can be accessed (C). A property inspector for the objects in the interface is displayed on the right (D).

The main document window is MainWindow.xib. To change the View Mode, use the three buttons in the upper-left corner. The buttons work much like they do in the Finder: the leftmost displays icons, the center a list, and the right one shows columns. The list view (which you can see in Figure 1-13), is more compact and easier to read (especially with longer names).

The Library window, at far left, contains a list of all the interface objects you can use in your design. You'll learn about these objects in detail in Chapter 3.

To start modifying the Flashlight's UI, double-click the Window item in the list in MainWindow.xib. The app's window opens to the right of the main document window and gives you the opportunity to modify the color. Start thinking about your favorite color!

At the far right is the Property Inspector. You'll use this window often as you refine your UI. Currently, it's showing Window Attributes because you're working with a window. The inspector is split into four main sections, which you choose from the tabs at the top. Each section lets you adjust various aspects of each object:

- **Attributes.** The object's particular settings. Behind the scenes, these items set properties and attributes for the object to save you from writing code (although you can still write the code manually if necessary).

- **Connections.** The connections define how your source code accesses the UI objects. For the Flashlight app, you'll see that "window" is connected to "Flashlight App Delegate".

- **Size.** This panel lets you define the selected object's geometry. For example, the window's width and height is 320 × 480 (the size of the iPhone's screen).

- **Identity.** This view shows what kind of object is defined. The app's window has a class of UIWindow.

Tip: To manage Interface Builder's many windows, you'll find that keyboard shortcuts make short work of finding the information you need. You can access the main document window by pressing ⌘-0, and you can switch to each section of the inspector with ⌘-1 through ⌘-4. Use Shift-⌘-L to bring up the Library window.

If you don't fully understand what's going on here, don't worry. You'll learn everything there is to know about classes, objects, and instances in the next chapter.

Note: If you've used other development environments, leave your preconceptions about how Interface Builder works at the door. Instead of automatically generating code to display the UI from resources, the .xib file contains an XML representation of the actual object instances and hierarchy. When you load a file, Interface Builder creates an object graph in memory and connects it to the instance variables that you've chosen. We'll go into greater detail about how this works in Chapter 3.

Now that you've had a little tour of Interface Builder, you can modify your flashlight's UI. Make sure that Window is selected in the MainWindow.xib document and that the Property Inspector is on the first panel (for modifying attributes).

Have you picked your favorite color yet? Time to change the background color of the window by clicking the Background color picker. Then click a nice color for the light in the color wheel (Figure 1-14). A soothing shade of yellow, for example.

The window preview updates immediately as you select colors. As your UI develops, you'll really appreciate this quick feedback: you don't have to build your app to see how a change will look, and that saves a lot of time.

Choose File→Save to update your MainWindow.xib file. Get in the habit of saving after tweaking your UI, and remember to check for the "modified" dot in the document's close box if your changes aren't working right. The last thing you want is to waste time debugging a user interface bug because you forgot to save the file.

Now switch back to the Xcode project window for the Flashlight, and select Build→Build and Run.

Note: If you see a window pop up with a warning to Stop Executable, just click OK. Xcode is just reminding you that you already have an application running in the simulator.

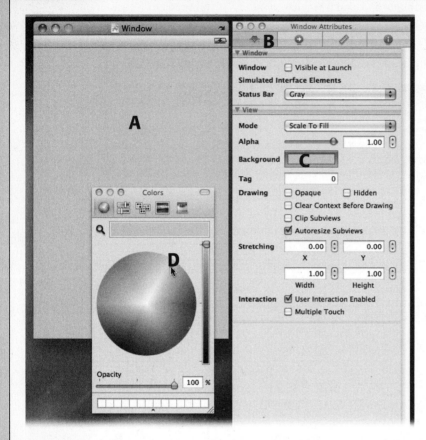

Figure 1-14:
Changing your application window's background color involves clicking a color wheel. After selecting the Window in the MainWindow. xib document (A), you view the attributes (B), which include a background color. Clicking in the well (C) brings up a color wheel (D) where you can select a new color with your mouse. As you select new colors, you get immediate feedback.

You've just made your first iPhone application and customized it to fit your own tastes. Well done!

Now that the journey is complete, read on to look at some details for things you saw along the way.

The Power of Brackets

You've now seen how easy it is to build an iPhone app: It's not rocket science. Or is it?

Apple has done an excellent job of creating the tools you use for building iPhone applications, but if you're going to become a master craftsperson, you need to learn more about these new tools. The first section of this chapter will get you started in this exploration: You're going to take a look the basics of the Objective-C language used to create apps.

Also, a thriving community of developers is working on the iPhone. You'll learn where to look for help with your coding questions, how to keep up with the ever evolving technologies used in the iPhone, and where to download free sample code to use in your own app development.

Note: There's a lot to learn in Objective-C. The book you're holding now would weigh several more pounds if the language were covered in depth. Instead, this chapter will cover the topics that you're most likely to encounter in iPhone development.

At the end of the chapter, you'll see some recommendations on where to look for additional information.

Once you've gotten up to speed with the essential parts of the language, you'll take a quick look at the documentation viewer that's built into Xcode. Knowing how to find information quickly and effectively will ease your entrée into this new platform.

Objective-C: The Nuts and Bolts for Your iPhone App

In this section, you'll explore the language used for programming iPhone applications: Objective-C. This chapter won't teach you how to program, but rather will show you the differences between this language and others you might have used. You'll also encounter features that are exclusive to Objective-C.

The Land of Square Brackets

While you were creating your flashlight, you might have noticed this line of code:

```
[window makeKeyAndVisible];
```

If you've programmed in Java, C, or JavaScript, those square brackets probably made you scratch your head. You're not the first developer to react that way!

So what are those funny brackets for? Before that question is answered, take a look at the language's name—the *C* on the end. You see, this whole new language is really based on an old one: the C language developed by Dennis Ritchie at Bell Telephone Laboratories in 1972. As with Java and JavaScript, the creators of Objective-C built something new on top of something familiar to many programmers. Once you learn a few tricks, this language will quickly become familiar to you, too (For more information on the relationship between Objective-C and other C languages, see the box below).

UNDER THE HOOD

C Plus Stuff

Objective-C is a superset of the C language syntax. Everything you see in this chapter is built using the GNU C/C++ compiler.

When a source file uses .m rather than .c, additional processing on the file occurs (similar, in concept, to using a preprocessor). This step hooks your code into a small runtime that supports the features of Objective-C. Classes and objects are simply C structures; the runtime lets them communicate with each other via introspection and message passing.

Because the entire infrastructure is based on standard C, it's easy to mix in open source libraries and legacy code with your Objective-C application. You don't have to give up your prized C/C++ source code when starting with this new language.

That said, many developers find working in this object-oriented environment quite addictive. It's fairly common to see Objective-C used as a wrapper that makes an open source library easier to work with. A good example is with regular expressions: Cocoa Touch has no native support for them. Thanks to the work of John Engelhart, the RegexKit Framework *http://regexkit.sourceforge.net/* makes Perl Compatible Regular Expressions (PCRE) a joy to use with your string objects.

After you've had a chance to get familiar with the syntax and feel of this new language, you'll explore how standard C idioms are used to implement it in the box on page 60.

Take a look at the following Objective-C code. You'll immediately feel comfortable with the familiar expressions and control structures you've used in other languages:

```
int i;
for (i = 0; i < 10; i++) {
    // check for even and odd values
    if ((i % 2) == 0)
        [label setTitle:@"Even"];
    else
        [label setTitle:@"Odd"];
}
```

In this contrived example, it's easy to see that you're checking for even and odd values in a loop that executes 10 times.

The first part of the language's name gives you a hint as to what those brackets are used for: *Objective* tells you that you're dealing with objects. And those square brackets let you communicate with those objects. In the first example, you were communicating with a *window* object. The second example interacts with a *label* object.

The Object of It All

So now you're probably thinking, "Great, but what's an object?" Objective-C is what computer scientists call an *object-oriented* language. You're going to be doing object-oriented programming as you develop your iPhone application.

Tip: If you're new to object-oriented programming, take a moment to read the Wikipedia page on this topic. The overview and history will help you come up to speed on the design motivations and terminology used in this style of programming: *http://en.wikipedia.org/wiki/Object-oriented_programming*.

Objects are chunks of memory that contain code and data for your application. One of the main tenets of object-oriented design is that your application should know as little as possible about the inner workings of an object. Terms like *encapsulation* and *data abstraction* describe this behavior.

An object should be like a black box. In fact, it may be helpful for you to use those square brackets as a visual reminder: [] looks a lot like a box.

As soon as you turn on your iPhone, thousands of objects are created in memory. Thousands more are added as soon as you launch your own application. You'll often hear *instance* used to describe each of these objects. With so many instances floating around, you'll need variables to keep track of the important ones.

Telling Your Objects to Do Things

The variables that reference objects let you change an object's behavior or state. You do this by sending a *message* to the object. In the Cocoa documentation, the object that's getting the message is often referred to as the *receiver*.

Note: You'll also see the term *target* used to represent the object that's getting the message. The message itself is sometimes called an *action,* since messages often cause things to happen. You'll see these terms often when dealing with user interfaces (in Interface Builder).

For the *window* object shown at the beginning of this section, you sent a message to *–makeKeyAndVisible.* In essence, the message is telling the window to make itself visible onscreen. Your "label" variable was sent the message *–setTitle:* along with a variable; it's being instructed to change its title to a new value. Don't worry about the action that takes place—your main focus right now is to understand how actions are initiated using messages.

As you get used to this new programming syntax, the square brackets remind you that you're communicating with a black box. The first word specifies the object that's receiving the messages, and the remaining words supply any other parameters that are needed. Here are some examples:

```
[myBlackBox messageWithNoParameters];
[myBlackBox messageWithFirstParameter:one andSecondParameter:two];
```

Note: If you're coming from a background in another object-oriented language, such as C++ or Java, sending messages is functionally equivalent to this:

```
myBlackBox->functionWithNoParameters();
myBlackBox->functionWithParameters(one, two);
```
or this:
```
myBlackBox.functionWithNoParameters();
myBlackBox.functionWithParameters(one, two);
```
The big difference is that the messages you send to objects can be generated at runtime as well as when your app is compiled. Similarly, object methods can be modified while your code is running. Objective-C is a dynamic language, so you can change the behavior of your objects at any time.

Each of these messages is handled by a *method* defined by the object. A method uses a unique signature that lets Objective-C route your message and any parameters to the right code for execution. You'll find that many method signatures are quite readable in your code: It's a little more typing, but you'll end up with code that's easier to understand.

Some methods return values. Earlier, you saw the *setTitle:* method change a label's title. There's a corresponding method called *title* that lets you query the object for the current value. For example:

```
NSString *labelTitle = [label title];
```

You're now thinking outside the black box. But what the heck is that *NSString* thing?

It's a variable definition. Just like any variable definition in C, the *labelTitle* variable is a pointer (because of the asterisk) to a class called *NSString*. That's great, but what's a class? Read on.

Masses of Classes

Every object in Objective-C belongs to a *class*, because that's where behavior and internal data are specified. A class is where the methods (code) are defined and implemented. It's also where you define what instance variables (data) are managed by the class.

NSString is a class that provides a rich set of methods for storing and manipulating text. The *NS* in the name means it's a part of a system foundation framework in Cocoa Touch.

Note: Frameworks are collections of classes and other resources you can add to your application. You'll learn all about the Cocoa Touch framework that supplies all the interesting classes, like *NSString,* in the next chapter.

Also, since Objective-C lacks namespaces, frameworks typically use a two-character prefix to prevent class name collisions. Foundation classes, which were originally written for NeXTSTEP, use *NS.* The user interface classes in Cocoa Touch use *UI.*

Some developers use their own prefix for class names to avoid these types of conflicts. For example, the Iconfactory uses *IF* for all classes that are shared among projects.

Therefore, *labelTitle* points to an instance of a string object that's defined by the system.

Because you have this object pointer in a variable, you can send messages to it. For example, you can send the *uppercaseString* message to your *labelTitle,* and it will return a new string object that contains all the lowercase letters converted to uppercase letters:

```
NSString *moreAwesomeLabelTitle = [labelTitle uppercaseString];
```

And that *moreAwesomeLabelTitle* variable, in turn, can be used to update the title on the original label:

```
[label setTitle:moreAwesomeLabelTitle];
```

Tip: Nesting objects and their messages can be an effective way to shorten your code and reduce the number of temporary variables. You can collapse these three lines of code:

```
NSString *labelTitle = [label title];

NSString *moreAwesomeLabelTitle = [labelTitle uppercaseString];

[label setTitle:moreAwesomeLabelTitle];
```

into a single line of code by using this technique:

```
[label setTitle:[[label title] uppercaseString]];
```

Just make sure to balance the brackets correctly or the compiler will complain!

You might have wondered earlier what the @*"Even"* and @*"Odd"* parameters were for in the *setTitle:* method on the label object. Good catch!

Since strings are used a lot in programming, the creators of Objective-C defined a shortcut for these text objects. Without the shortcut, @*"Even"* would need to be instantiated using:

```
[NSString stringWithUTF8String:"Even"]
```

That's a lot more typing. And the best programmers are the lazy ones. Welcome to the club!

Classes in Detail

At this point, it should be pretty obvious that classes are really important in Objective-C. They act as building blocks for your application, and they let you avoid writing a lot of boring code. It's not hard to envision how an *uppercaseString* method would be implemented, but it's even better to not think about it at all and to just let the system handle the work.

Classes are defined in header files that are included in your source code. System-level classes, like *NSString,* get incorporated by default. Take a look at the top of the FlashlightAppDelegate.h file that was created automatically for you in the last chapter:

```
#import <UIKit/UIKit.h>
```

That one line of code pulls in the class definitions for the entire system. The pre-processor begins by loading the classes for developing Cocoa Touch user interfaces (*UIKit*). These classes, in turn, load foundation classes for managing data and the core graphics classes for drawing.

Note: For you experienced C programmers, note that there's a subtle difference between #*include* and #*import*: The preprocessor will keep track of imports and not include them more than once. This technique makes it much easier to avoid multiple definitions!

What do these header files for classes look like? Here's a part of the class definition for *NSString* taken from the NSString.h header file:

```
@interface NSString : NSObject <NSCopying, NSMutableCopying, NSCoding>
- (NSUInteger)length;
- (unichar)characterAtIndex:(NSUInteger)index;
@end
```

Even if you have experience working with other programming languages, this code won't make any sense. Time to deconstruct the syntax!

First there's *@interface*:

```
@interface NSString : NSObject <NSCopying, NSMutableCopying, NSCoding>
```

This code tells Objective-C that you're about to define data and methods for a class. This, in effect, is a template for every object that gets instantiated from the class. The interface is also where the class name is defined as *NSString*.

The interface definition continues until the *@end* and the, well, end.

One class to rule them all

After definition, there's a colon and some other names beginning with *NS*. That's your hint that you're dealing with system class definitions. The most important one is *NSObject*:

```
@interface NSString : NSObject <NSCopying, NSMutableCopying, NSCoding>
```

All classes, and therefore objects, in Objective-C "inherit" from another class. *Inheritance* lets the characteristics of a parent class pass on to a child class. The child class—more commonly referred to as a *subclass*—can then modify the methods and instance variables that were defined by the parent. (The parent is often called a *superclass*.)

Tip: There's one exception to this inheritance rule, and that's *NSObject*. It's the root class in the hierarchy, and it doesn't inherit from any other class.

Much as with your parents, inheritance gets you a lot of behavior for free (even if you don't want or need it!). Even if there's no money involved with this inheritance, you'll find it invaluable because it saves you a lot of time and effort. (For details on this "free code," see the box on page 36.)

An example of inheritance occurs in custom views. It's likely that at some point in your development, you'll want to develop a custom view for displaying some application-specific data. You'll start this work by subclassing a system class like *UIView*.

As you implement your child class of *UIView*, you'll find that a lot of code has already been done for you. Things like handling multitouch events, drawing, animation, and view management are available from the parent class. You only have to implement the new behavior for the view.

Now that you've learned a bit about the class hierarchy in Objective-C, it should be clear that the first word after the colon is the superclass (parent) of the class being defined. *NSString*'s superclass is *NSObject*.

Follow the protocol

After the superclass name, you'll see some other class names in angle brackets (< >):

```
@interface NSString : NSObject <NSCopying, NSMutableCopying, NSCoding>
```

These are *protocols* that are adopted by the class. Protocols define a group of methods that aren't associated with any particular class.

Using protocols is like signing a contract. If you specify that your class adopts a protocol by including the class names in the angle brackets, you promise to implement the methods that are defined in the protocol. For example, the *NSString* protocol promises to fulfill the contract of providing copies of itself. These object copies can be read-only (with *NSCopying*) and writable (with *NSMutableCopying*). The *NSCoding* protocol tells you that the class also implements methods for encoding and decoding objects.

POWER USERS' CLINIC

Tell Us about Yourself

The concept of free code extends all the way through the object hierarchy. As you're working on a *UIView* subclass, you'll have access to the code from *UIView*, *UIResponder* (where touch events are handled), and *NSObject* (the root class).

To see the power of the hierarchy, look at the *–description* method in *NSObject*. Since every object is a subclass of this root class, they all know how to describe themselves. The default implementation of the method returns the object's class name and a memory address. So even if you don't know anything about an object, you can do this:

```
[mysteryObject description]
```

The resulting string that describes the *mysteryObject* is helpful when you're debugging or working with data from a source outside of your direct control.

For example, the debugger that you use with Xcode (gdb) includes a *print-object* command. When you type this into the debugger:

```
(gdb) po myObjectVariable
```

a description message is sent to the object, and the result is then printed on the console.

This behavior can also be handy when you're logging information. Cocoa's logging function, *NSLog,* takes a *printf*-style string and outputs the results to the current console device:

```
NSLog(@"THIS OBJECT IS AWESOME: %@",
    chockLockObject);
```

The formatting specification is supplied with an *NSString* (remember the @ shortcut mentioned earlier). There's also a new formatting specification—%@—that prints an *NSString* just as %s prints a C-style string.

The really cool part is that the logging function is also smart enough to know that *chockLockObject* isn't an instance of *NSString,* so it uses the object's description method to create the string that's displayed.

Note: The encoding and decoding methods in *NSCoding* are used to archive objects on disk or to distribute them across a network. It's Cocoa's mechanism for serializing objects.

Protocols aren't required when you're defining a new class, but you'll find them used quite a lot as you develop your app. The most common use of protocols is with delegates; you'll see that design pattern in the next chapter.

Note: Since Objective-C supports only single inheritance, protocols are often used to achieve the same goals as multiple inheritance in other languages.

The Methods Behind the Madness

That's a lot of reading just to understand a single line of code! Thankfully, it gets a little easier with the next two lines, both of which begin with a minus sign (–).

```
- (NSUInteger)length;
- (unichar)characterAtIndex:(NSUInteger)index;
```

This is how you define methods (page 49) in a class interface. You can represent any string in Cocoa with just these two methods: The first returns the length of the string; the other returns a Unicode character at an index position.

Tip: When writing about methods in online forums, mailing lists, or blogs, many developers use an abbreviated form for the signatures: They leave out type information and parameter names. Apple's developer documentation also uses this format.

It's common to see something like "I'm having problems with *–messageWithFirstParameter:andSecond Parameter:*" in a post asking for help. Another reason to keep those method signatures readable! The short forms for the *NSString* methods shown above are *–length* and *–characterAtIndex:*. This book uses this format throughout.

The definitions also include types for any parameters and the result. The *–length* method returns an unsigned integer (*NSUInteger*). The *–characterAtIndex:* method takes one parameter with an unsigned integer index and returns a unichar. These methods use primitive types, but you can also specify objects using a pointer to a class name (you'll see that shortly with the *–uppercaseString* definition).

Tip: The types used in the method definition are checked at compile time, but not when messages are sent at runtime. When you see a compiler warning that "'NSString' may not respond to '–aNonexistent-Method'", make sure that you're sending the message to the right kind of object.

When the compiler warns about "Passing arguments" from a method, that's your clue that you have mismatched types in the message you're sending to an object. If you see a message about incompatible pointer types, make sure you're using @"*string*" instead of a standard C "*string*". The "without a cast" warnings are usually an indicator that you've used a primitive type instead of an object (or vice versa).

If you ignore these warnings, you'll still be able to run your app, but it will crash as soon as these incompatible types are sent in a message.

Note that this is just a class definition, so there's no mention of where or how your string data is stored. Remember that you don't really need to know about the inner workings of an object. (You'll learn to love this encapsulation thing.)

But you may be asking yourself, "What happened to that *–uppercaseString* method I saw used on page 33?" That's a good question, and it leads into another powerful, and unique, feature of Objective-C: categories.

Categorically Speaking

One of the problems with important classes—like the ones for handling strings—is that they can grow quite large and unwieldy. Categories help avoid this by breaking the class definition into pieces. In the case of *NSString*, only the most basic functions are present in the main class definition. The real meat of the class is in this category:

```
@interface NSString (NSStringExtensionMethods)
```

There's the *@interface* you saw earlier (page 35), followed by the class name. It's only when you get to the parentheses that things get interesting.

The parenthesized name gives you a hint as to what's going on here. This class interface supplies methods that extend the previously defined *NSString*. It's adding functionality without cluttering up the basic definition of a string.

Tip: There's an art to naming in Objective-C. It's OK to be wordy, as with the *NSStringExtensionMethods* category name. Just make sure the name is descriptive enough that you can understand it without referring to documentation.

This string category contains the following method definition:

```
- (NSString *)uppercaseString;
```

That's the exact method you saw on page 33.

But categories get even cooler. You can use them to extend classes that already exist (and that you don't even have the source code to). Imagine you need to make all your strings *awesome*. You don't really know what awesome means yet, but you do know that you'll need to do it often.

You start by defining your own category like this:

```
@interface NSString (AwesomeMethods)
- (NSString *)awesomeString;
@end
```

You're basing your category on the existing class *NSString,* and you assign it a category name of *AwesomeMethods* by including it between the parentheses.

The new class will have all the capabilities of a normal *NSString,* but it will also respond to one additional message named *–awesomeString.* But it gets even better: All other string objects in the system will understand the new message. If you had implemented *–awesomeString* in a subclass, you'd spend a lot of time refactoring your own code to use the new implementation, and have issues with code out of your control (such as strings returned by a Cocoa API). Categories are a powerful construct that lets you extend the system without breaking what is already there.

All you need now is some code to make those awesome strings.

Implementation: The Brains Behind the Beauty

For every *@interface,* there's an *@implementation.* The easiest way to think about it is that *@interfaces* are what your class looks like from the outside, and the *@implementation* is what it looks like from the inside.

Note: So far, you've been looking at what goes in an interface header file. These files, like in C, use the .h extension by convention. An interface's implementation goes in a .m file. Just remember that *M*ethods go in a file that ends with the letter *M.*

Here's what the implementation could look like:

```
@implementation NSString (AwesomeMethods)

- (NSString *)awesomeString {
    NSString *awesome = [self uppercaseString];
    if (! [self hasSuffix:@"!"]) {
        // ADD SOME IMPACT!!!
        awesome = [awesome stringByAppendingString:@"!!!"];
    }
    return awesome;
}

@end
```

By now, you're getting better at reading Objective-C code. You'll see that an awesome string is created using your old friend *–uppercaseString.* The *–hasSuffix:* method checks to make sure that there's at least one exclamation point at the end. If not, the *–stringByAppendingString:* method adds the necessary impact. But what does that *self* mean?

Every method implementation is passed a hidden argument named *self* that references the object receiving the message. Since your method is a part of the *NSString* class, *self* refers to an instance of that class. (Yes, you're talking to yourself in this method, and that's OK.)

Note: C++, Java, and PHP all use *this* to provide this self-reference.

Once you've implemented this category method, your code can call it like this:

```
NSString *ordinaryString = @"typing power";
NSString *excitingString = [ordinaryString awesomeString];
```

The *excitingString* object will have the value "TYPING POWER!!!".

Categories also help you avoid repetition in your own code. Suppose you do a lot of checking of strings to see if they're awesome. You'd be writing code like this over and over again:

```
NSString *myString = @"something";
if ([myString isEqualToString:[myString awesomeString]]) {
    // sorry, but myString isn't awesome
}


NSString *myOtherString = @"SOMETHING ELSE!!!";
if ([myOtherString isEqualToString:[myOtherString awesomeString]]) {
    // myOtherString, on the other hand, is truly awesome
}
```

That's a pain to type and a pain to read. So add this to your *NSString*'s category interface:

```
@interface NSString (AwesomeMethods)
- (NSString *)awesomeString;
- (BOOL)isAwesomeString;
@end
```

And add this to its implementation:

```
- (BOOL) isAwesomeString {
    return [self isEqualToString:[self awesomeString]];
}
```

This cleans up your checking code quite a bit:

```
NSString *myString = @"something";
if ([myString isAwesomeString]) {
```

```
        // sorry, but myString isn't awesome
    }

    NSString *myOtherString = @"SOMETHING ELSE!!!";
    if ([myOtherString isAwesomeString]) {
        // myOtherString, on the other hand, is truly awesome
    }
```

It's also important to note that at this point you haven't actually created a new class. You've only extended code provided in a system framework to suit your own needs. Cocoa Touch is awesome; you just made it more awesome.

Creating New Classes

As you can see, categories can accomplish a lot and provide a great way to extend code that you didn't write. But they have one major limitation: You can't add instance variables in your category definition. And with the *NSString* class, how could you? The class interface tells you nothing about how the string is stored.

It's time to learn how to create new classes and to extend the Cocoa hierarchy. And to see how that's done, you're going to create a new class that lets you control the number of exclamation points, as in the phrase "THIS IS GOING TO BE AWESOME!!!!!!!"

The first thing to do is create an *@interface* for this new class:

```
@interface AwesomeStringMaker : NSObject
{
    NSNumber *exclamationCount;
    NSString *originalString;
}
- (NSNumber *)exclamationCount;
- (void)setExclamationCount:(NSNumber *)newExclamationCount;
- (NSString *)originalString;
- (void)setOriginalString:(NSString *)newOriginalString;

- (NSString *)awesomeString;
@end
```

This code looks similar to the class category on page 39. This time, the category name (in parentheses) is gone, and some new code is in curly braces: { }. A class's data is specified between these braces.

This example has two instance variables: a number that keeps track of the number of exclamation points and a string that you want to make awesome.

Note: You'll find that class data is referred to in many ways. Objective-C developers use the terms *instance variable, ivar,* and *property* interchangeably. If you're talking to a longtime C++ developer, you may hear *member variables.* Java diehards use *fields.*

Whichever term is used doesn't matter; it's still just a block of memory that's associated with each instance of your class.

A few new methods at the end of the interface let you read and write the data managed by this new class. These methods are called *accessors* because they let you access the class's internal information. By convention, the method that reads the instance variable just uses the name. The method that updates the instance variable is prefixed with *set.* Other languages call these methods *getters* and *setters.*

Only the implementation of this class has easy access to all instance data. Without the accessor methods, the data is essentially hidden. In most cases, that's a good thing, especially when you have private data that you don't want exposed to the caller.

Now for the *@implementation* of your *AwesomeStringMaker* class:

```
@implementation AwesomeStringMaker

- (NSNumber *)exclamationCount {
    return exclamationCount;
}

- (void)setExclamationCount:(NSNumber *)newExclamationCount {
    if (exclamationCount != newExclamationCount) {
        [exclamationCount release];
        exclamationCount = [newExclamationCount retain];
    }
}

- (NSString *)originalString
{
    return originalString;
}

- (void)setOriginalString:(NSString *)newOriginalString {
    if (originalString != newOriginalString) {
        [originalString release];
        originalString = [newOriginalString copy];
    }
}

- (NSString *)awesomeString {
    NSString *awesome = [originalString uppercaseString];
    NSUInteger length = [awesome length];
```

```
    NSInteger padding = [exclamationCount unsignedIntValue];
    NSString *moreAwesome = [awesome stringByPaddingToLength:(length +
    padding) withString:@"!" startingAtIndex:0];

    return moreAwesome;
}

@end
```

And with this *AwesomeStringMaker* power, you can use the class like this:

```
AwesomeStringMaker *myAwesomeStringMaker = [[AwesomeStringMaker alloc] init];
[myAwesomeStringMaker setExclamationCount:[NSNumber numberWithFloat:8.0f]];
[myAwesomeStringMaker setOriginalString:@"typing power"];

NSString *myAwesomeString = [myAwesomeStringMaker awesomeString];

// myAwesomeString now contains "TYPING POWER!!!!!!!!"

[myAwesomeStringMaker release];
```

Tip: As you're learning Objective-C, be careful which objects get which message. If you try to send a
–*setExclamationCount:* message to an instance of *NSString* instead of to *AwesomeStringMaker,* you'll get
a "method not found" warning while compiling your code. At runtime, your application will crash with an
exception in *objc_msgSend* because the *NSString* class doesn't know what to do with the message you've
sent.

Managing Memory

If you take a look at your class's accessors for *exclamationCount* and *originalString,*
you'll see some new methods being called on the instance variables. The *retain, copy,*
and *release* messages are extremely important: You use them to manage the memory
usage of objects.

Note: Memory is also one of the more complicated things to understand about Objective-C, so don't get
frustrated if you don't understand it at first.

As you learned before, thousands of objects are created and deleted every second
that you're using your iPhone. And each of those objects uses a chunk of memory,
which is a precious resource on a mobile device. Your desktop computer may measure its memory capacity in gigabytes, but your phone only has a few hundred
megabytes.

If you're not careful about your memory usage, the iPhone's operating system will shut down your application. If it didn't, the system would eventually grind to a halt, and you'd have to reboot your phone. That's kind of inconvenient when you're expecting an important call.

So how is memory managed with these *–retain, –release,* and *–copy* methods?

Every object in memory maintains a counter. This counter keeps track of how many other objects are using the object. When you want to use an object, you must tell the object, so it can update its counter, and you do so by using the *–retain* message. After you send that message, the object updates its retain count, and that object won't be deleted. If you're ever interested in how many other objects are keeping a reference to that object, you can send it the *–retainCount* message, and it'll return the current value of the counter. When objects are initially allocated, their retain counter is set to *1.*

Now that you know how you keep objects around in memory, you need a way to get rid of them when you're finished using them. That's where the *–release* message comes in. It does the exact opposite of *–retain*; instead of incrementing, it *de*crements the retain count. You're telling the object you don't need it anymore.

As your app runs in the Cocoa Touch environment, the retain count of objects is checked periodically. Objects with retain counts that have reached zero aren't being used by any other objects, so the system can delete them and reclaim their memory.

Now that you're familiar with the mechanics of retain and release, take a look at the accessor in more detail:

```
if (exclamationCount != newExclamationCount) {
```

This line is a simple performance optimization. If the old object and new object are the same, there's no need to adjust the retain count.

Tip: This code is comparing pointer addresses, which is a quick (and valid) way to check that two objects are the same. In fact, *NSObject's –isEqual* method uses this same method. (For many classes, such as *NSString,* checking for equality is more complicated than checking pointers.)

The first thing to do is tell the old instance variable that you no longer need it by sending the *–release* message:

```
[exclamationCount release];
```

Now that you've released the instance variable, you have no guarantee that the variable is valid. If you try to send additional messages to *exclamationCount,* your app is likely to crash. So the next step is to retain the new object that is being passed into the accessor. The retain method returns itself, so that value is used to update the instance variable for *exclamationCount*. At this point, it's safe to send messages to this object again:

```
exclamationCount = [newExclamationCount retain];
```

A variation on the *–retain* message is the *–copy* message. As with retain, you'll have a unique reference to an object that won't go away until you decide that it's OK. The difference is that you'll be holding onto a copy of the original object's data.

Here's a case where *–copy* comes in handy. With the *originalString* instance variable, you don't want to use *–retain*, because the original string could be modified by another object. Remember, if you're sharing a reference with many other objects and one of them changes the string, you'll see the change when you call *–awesomeString*.

You want to have your own unique copy of the string, so you use this instead:

```
originalString = [newOriginalString copy];
```

If another object modifies the object pointed to by *newOriginalString*, you won't be affected.

Take a *nil* Pill

Sometimes you want to clear out an instance variable completely. You might want to save some memory and remove the object because you're no longer using it. Or you might be modeling some state within your application where the absence of an object is important. You can clear the variable with a special object called *nil*. If a variable referencing an object contains the value *nil*, there's no object.

Nil objects have an interesting behavior that affects how you use them while managing memory. When you send a message to a *nil* object, the message is ignored and a result of *nil* is returned. This technique is commonly called a *nil targeted message* and it looks like this:

```
NSString *missingString = nil;
NSString *excitingString = [missingString awesomeString];
// excitingString has a value of nil, too
```

When you're writing accessors with *–retain*, *–copy*, and *–release*, this feature lets memory get cleaned up. To see how this works, replace *newExclamationCount* with *nil*, and you have this code:

```
if (exclamationCount != nil) {
    [exclamationCount release];
    exclamationCount = [nil retain];
}
```

If the current count isn't already *nil*, the object for *exclamationCount* is released. Then the *nil* result of the retain message is used to set the new count.

Nil targeted messages can also help prevent crashes in code that has failed some kind of initialization. (Objects that fail to initialize properly are assigned *nil* to show that they don't exist.) You can also use *nil* objects to your advantage for those cases where you want to manage a state. An example would be a *middleName* instance variable in a view: If the value of the variable is *nil*, you know that you don't need to display a person's middle name.

Some code will also take advantage of *nil*'s value: zero. In the case of the *exclamationCount*, when it's *nil*, the message for *unsignedIntValue* is ignored and *nil* is returned:

```
NSInteger padding = [nil unsignedIntValue];
// padding has a value of zero
```

With a padding of 0, no exclamation points will be used, which seems like the right behavior when there's no object for the instance variable. On the other hand, if your code needs an object but one has not been set, you'll want to use something like this:

```
if (myObject) {
    [myObject doSomething];
}
else {
    NSLog(@"Can't doSomething because myObject is nil!");
}
```

Again, this code relies on the fact that a *nil* object has a value of zero. If there's a value for the *myObject*, its nonzero value will let the *–doSomething* message be sent.

Tip: *Nil* objects can also be a great source of head scratching. If you forget to initialize a variable, you'll wonder why an object instance isn't doing what you expect when you send it messages.

The "aha" moment will come when you go into the debugger, display the object, and see "Cannot access memory at address 0x0". Your object's value is zero.

Autorelease with Ease

Autorelease has nothing to do with letting go of your car. All of this retaining, copying, and releasing is a lot of cumbersome programming when you're dealing with temporary objects. Often, you'll want to keep an object around just long enough to do some work. For example, many objects never exist outside of the scope of a method implementation.

The brilliant and lazy programmers who came up with Objective-C have a workaround called the *–autorelease* method. To see how it works, first look at how the *myAwesomeStringMaker* object is maintained:

```
AwesomeStringMaker *myAwesomeStringMaker = [[AwesomeStringMaker alloc] init];
```

```
// do some stuff with myAwesomeStringMaker
```

```
[myAwesomeStringMaker release];
```

This code can be simplified by using the *–autorelease* message:

```
AwesomeStringMaker *myAwesomeStringMaker = [[[AwesomeStringMaker alloc] init]
autorelease];
```

```
// do some stuff with myAwesomeStringMaker
```

After the object has been allocated and initialized, it's sent the *–autorelease* message. An object that's been autoreleased is guaranteed to remain in memory for the current method. After that, it will be deleted without your direct intervention.

This phenomenon makes things much easier for you. In this example, *myAwesome-StringMaker* is only needed for a short time, so autorelease does the object cleanup automatically. If you're doing a lot of things with *myAwesomeStringMaker* between allocating the object and the end of the method, it's easy to forget to send the message to *–release.*

Because other programmers are just as smart and lazy as you are, many methods return autoreleased objects. The assumption is that you're probably not going to need that object for a long time, so it's best to let the system do the cleanup.

By convention, only methods that have *init* or *copy* in their name return objects that aren't autoreleased. If you use one of these methods, you've got to pay attention to what's happening with your memory. See the box below for more detail.

WORD TO THE WISE

Memorize the Rules

Longtime Cocoa programmers will cite the following rules, originally written by Don Yacktman, whenever you ask a question about how to manage memory:

- If you allocated, copied, or retained an object, then you're responsible for releasing the object with either *–release* or *–autorelease* when you no longer need the object. If you did not allocate, copy, or retain an object, then you should not release it.

- When you receive an object (as the result of a method call), it will normally remain valid until the end of your method, and the object can be safely returned as a result of your method. If you need the object to live longer than this—for example, if you plan to store it in an instance variable—then you must either *–retain* or *–copy* the object.

- Use *–autorelease* rather than *–release* when you want to return an object but also wish to relinquish ownership of the same. Use *–release* wherever you can, for performance reasons.

- Use *–retain* and *–release* (or *–autorelease*) when you want to prevent an object from being destroyed as a side effect of the operations you're performing.

Properties and Dots

As you get more advanced with your Objective-C coding, you'll find that you start to have a lot more accessors in your classes. They're another case of boring code that the lazy programmer wants to avoid, especially all of the retain and release stuff that's repeated over and over. You can utilize another shortcut called *declared properties*. This language feature provides a simple way for you to specify, implement, and use a class's accessor methods.

Both the class interface and implementation change when you're using properties:

```
@interface AwesomeString : NSString
{
    NSNumber *exclamationCount;
    NSString *originalString;
}
@property (nonatomic, retain) NSNumber *exclamationCount;
@property (nonatomic, copy) NSString *originalString;

- (NSString *)awesomeString;
@end
```

Each property is configured using the attributes in parentheses. The *nonatomic* attribute makes the accessor faster; you only need atomic properties when working across multiple threads. The *retain* attribute causes the instance variable to use the retain/release pattern during assignment. Likewise, the *copy* attribute uses the copy/release pattern.

You just got rid of two lines of code, but wait, there's more!

In the implementation, you can replace the *exclamationCount* and *setExclamation-Count* methods with one line of code. And you can replace the *originalString* and *setOriginalString* with a second line of code:

```
@implementation AwesomeString

@synthesize exclamationCount;
@synthesize originalString;

…
```

The synthesize shortcut causes the compiler to generate the code necessary to access your instance variables. The code in the *@implementation* is generated according to the *@property* definition in the *@interface*. You may get tired of typing all those @ symbols, but those *@synthesize* statements saved you from writing 19 lines of code. With a larger class, the savings in lines of code will be even more impressive.

Once you've defined properties for your class, you may want to access them using *dot notation*. This notation lets you access the instance variables without using the square brackets. Finally, some relief for the C++ and Java coders amongst us! For many developers this code will be much more readable:

```
AwesomeStringMaker *myAwesomeStringMaker = [[[AwesomeStringMaker alloc] init]
autorelease];

// use dot notation to set a property's value...
myAwesomeStringMaker.exclamationCount = [NSNumber numberWithFloat:8.0];
myAwesomeStringMaker.originalString = @"typing power";
NSString *myAwesomeString = [myAwesomeStringMaker awesomeString];

// or use dot notation to read a property
if ([myAwesomeStringMaker.exclamationCount integerValue] < 4) {
    // that's not awesome enough
    myAwesomeStringMaker.exclamationCount = [NSNumber numberWithFloat:8.0];
}
```

A property on the left side of the equals sign will use the setter. The same property on the right side will use the getter.

Methods of Class

Earlier, you might have been a little confused by this code:

```
[NSString stringWithUTF8String:"Even"]
```

With all the talk of sending messages to object instances, now you're not using one. Where the heck is that message going?

To begin answering that question, take a look at the definition in the *NSString* interface:

```
+ (id)stringWithUTF8String:(const char *)bytes;
```

Yep, it's another wacky character courtesy of Objective-C. This time it's a plus sign (+).

On page 37, you saw that methods were defined with a minus sign at the beginning. But those are just *instance* methods. *Class* methods are another kind.

With instance methods, you need a variable that references the object instance before you can send the message. Sometimes this is a limitation in your class design: You'd like to make a method available without an object instance. Class methods are Objective-C's solution to this problem. With these methods, you send messages directly to the class without needing an actual object.

These class methods are often used to create new object instances. The *+stringWith UTF8String:* method is just such a case; sending the message to the class lets it return a new object for you to use in your code.

Note: You may be familiar with using *factory* objects. That's exactly what sending a message to the class does. The *+stringWithUTF8String:* message acted as a factory for a new instance of *NSString*.

So what could make your *AwesomeStringMaker* class more awesome? A class method that returns the *mostAwesomeString*:

```
@interface AwesomeStringMaker : NSObject

...


+ (NSString *)mostAwesomeString;
@end
```

Your implementation would look like this:

```
@implementation AwesomeStringMaker

...


+ (NSString *)mostAwesomeString {
    return @"CHOCKLOCK!!!!!!!!!";
}
@end
```

Now you don't even have to think when you need to make your app a lot more awesome:

```
[label setTitle:[AwesomeStringMaker mostAwesomeString]];
```

Initializing Objects

Your *AwesomeStringMaker* class has a bug. Hard to imagine, isn't it?

```
AwesomeStringMaker *myAwesomeStringMaker = [[[AwesomeStringMaker alloc] init]
autorelease];
myAwesomeStringMaker.originalString = @"typing power";
NSString *myAwesomeString = [myAwesomeStringMaker awesomeString];
```

The value for *myAwesomeString* is "TYPING POWER". It's missing some awesome exclamation points!!!!! That's because when a new instance of an object is created, all instance data is cleared out. Pointers are set to *nil* values and numbers are set to zero. This means your *exclamationCount* instance variable is zero. And you can't be awesome with zero exclamation points.

This situation illustrates a larger problem: Your objects usually need to be set up before you can send them messages. With Objective-C, there's a method for just that:

```
- (id)init {
    if (self = [super init]) {
        // WHY USE A LITTLE WHEN YOU CAN USE A LOT
        exclamationCount = [[NSNumber numberWithInt:8] retain];
        originalString = [@"" copy];
```

```
        }
        return self;
    }
```

That code may look convoluted, but what it does is simple and elegant: After every object is allocated in memory, the new instance is sent the *–init* message. You can choose to ignore the message, but it's more likely that you won't.

Note: You may be wondering why every object gets sent the *–init* message. It's because the default implementation is in *NSObject,* and every object is a descendent of this root class. If you don't implement the method, it will be handled by superclasses, including *NSObject.*

When your object gets the message, it assigns the result of sending *–init* to the parent object (represented by the implicit variable *super*) to the variable *self.* So what's all that about?

Just like your class needed to do some initialization, your parent class may need to do some. You don't know exactly what that setup is, but now is the only opportunity for any classes between you and *NSObject* to get their houses in order.

Once you've given everyone else a chance to get set up, you check the result of the assignment to *self.* It's possible that one of the superclasses wasn't able to initialize itself, so it returned *nil.* If that's the case, you'll skip your initialization and return *nil,* too.

Assuming that everything went fine with your superclasses, you'll get your chance to set the *exclamationCount* to a whopping 8 characters.

Deallocation Location

There's still a pretty serious bug in your code. You went to all the work of retaining, copying, and releasing your accessors. But what happens when the whole object is released? Unless you called each accessor and set a *nil* value, the memory used by instance variables will never be freed.

As you implement classes, it's your responsibility to clean up after yourself as objects of that class are released. The way you do this is with a *–dealloc* method. Just as *–init* let you get things set up, the *–dealloc* method lets you tear things down. Here's what that method should look like for your *AwesomeStringMaker* class:

```
- (void)dealloc {
    [exclamationCount release];
    [originalString release];

    [super dealloc];
}
```

First, both of your instance variables are released so that memory can be reclaimed. Then the *–dealloc* message is forwarded to the superclass so it can do its own cleanup.

Manual override

You may have noticed that *–init* and *–dealloc* were never defined in your *@interface*. That's because it's not necessary; as a descendent of *NSObject,* the method has already been declared. Your own implementations of these methods merely "override" those defined in the superclasses.

A class's documentation will give you guidance about which methods can be overridden. In some cases, it's required: If you create a *UIView* class that draws its own content, you must override the *–drawRect* method to display the view. In other cases, overriding methods provides flexibility and customization. For example, with the *–layoutSubviews* method, an override lets you explicitly control the positioning of interface elements in a way that you can't by using the default mechanisms.

Loops: For Better or For Worse

Looping with variables is probably one of the first programming concepts you learned. You'll be happy to know that Objective-C has *for* loops—and the syntax is better than that of plain C.

You'll learn all about the *NSArray* class in the next chapter, but for the moment all you need to know is that it's a class that manages zero or more objects. When you have more than one of something, it's likely that you'll want to loop over the contents at some point.

You could use the good old *for* loop in standard C:

```
NSArray *myArray = [NSArray arrayWithObjects:@"THIS", @"IS", @"AWESOME!!!",
nil];
NSUInteger count = [array count];
NSUInteger i;
for (i = 0; i < count; i++) {
    NSString *element = [myArray objectAtIndex:i];
    if ([element isAwesomeString]) {
        NSLog(@"CONGRATULATIONS!!!!!!!!!");
    }
}
```

But Objective-C has a construct called *fast enumeration* that saves a bunch of typing:

```
NSArray *myArray = [NSArray arrayWithObjects:@"THIS", @"is!", @"AWESOME!!!!",
nil];
for (NSString *element in myArray) {
    if ([element isAwesomeString]) {
        NSLog(@"CONGRATULATIONS!!!!!!!!!!");
    }
}
```

More importantly, this form of enumeration is optimized by the runtime and removes the possibility for bugs with loop invariants.

Another example of smart-but-lazy code that's simpler, safer, and faster.

Your Exceptional Code

As a developer, you know that things don't always go the way you plan. Code that doesn't work like you expect is a never-ending source of joy in your life. Fortunately, Objective-C supplies compiler directives that help you deal with exceptions in blocks of code:

- Use *@try* for code that may throw an exception.
- *@catch()* lets you specify code that gets run when exceptions occur. You can use more than one block if you need to handle different types of exceptions.
- You can use *@finally* if you have code that needs to run whether or not an exception occurred.

Here's an example that uses all three:

```
NSArray *myArray = [NSArray array]; // an array with no elements
@try {
    // the array is empty, so this will fail:
    [myArray objectAtIndex:0];
}
@catch (NSException *exception)
{
    NSLog(@"name = %@, reason = %@", [exception name], [exception reason]);
}
@finally
{
    NSLog(@"Glad that's over with...");
}
```

which will generate the following output in the console log:

```
name = NSRangeException, reason = *** -[NSCFArray objectAtIndex:]: index (0)
beyond bounds (0)
Glad that's over with...
```

If you hadn't caught the exception, your application would have quit and sent the user abruptly to the iPhone's home screen. Catching exceptions is a good thing.

When your code needs to generate an exception, use the *NSException* class. Suppose you come across a string that's not awesome enough; you could raise an exception like this:

```
[[NSException exceptionWithName:@"AwesomeException" reason:@"Not awesome
enough" userInfo:nil] raise];
```

You can use the *userInfo* parameter if you need to supply some additional information along with the exception (the offending string, for example).

Tip: When debugging, it's often helpful to set a breakpoint just before the exception is thrown. This way, you can see the stack trace at the point where the error occurred. Add a breakpoint at *objc_exception_throw,* and the debugger will show you everything that led up to the problem.

Learn by Crashing

It may seem odd that a chapter that's explaining a programming language includes a section on using the debugger. But think about the last time you learned a new language: You had plenty of bugs, because everything was new and you were doing things a more experienced developer would avoid.

Xcode has a powerful debugger built-in; it's based on *gdb* from the GNU development tools. Like with the Objective-C compiler, some extensions to the debugger make it compatible with objects and the runtime architecture.

When you run your code, you can do so with or without breakpoints enabled. Xcode's toolbar has a button that toggles your breakpoints: You can also choose Run→Debug→Breakpoints On (or press ⌘-Option-Y) to launch your app with them enabled.

Once you have breakpoints enabled, you can set new ones in a couple of ways:

- When you click in the left gutter of your source code, a blue arrow will appear (Figure 2-1). The arrow signifies that a breakpoint is set and that your application will stop when it's reached. If you click the arrow again, it becomes dimmed and disabled. You can also Control- or right-click the arrow to get a context menu with other options.

- All of the current breakpoints are displayed when you choose Run→ Show→Breakpoints (or press ⌘-Option-B). You can add a breakpoint to the list that's displayed by double-clicking the last line and entering the symbolic name. You'll need to use this method when you want to set a breakpoint where you don't have the source code (for example, in the Cocoa Touch frameworks). You can also remove a breakpoint in this window by selecting it and pressing Delete.

When setting breakpoints manually, you can use a function name such as *objc_exception_throw,* and the debugger will halt before it's executed. But since you're new to Objective-C, it's more likely that you'll want to break at a class method. Here's some special syntax to set these types of breakpoints:

- To stop at an instance method, use the form *–[ClassName methodName:].* For example, to stop in *NSArray's –objectAtIndex:* method, you'd use *–[NSArray objectAtIndex:].* (Make sure to remember the semicolons when specifying the name, or the breakpoint won't fire.)

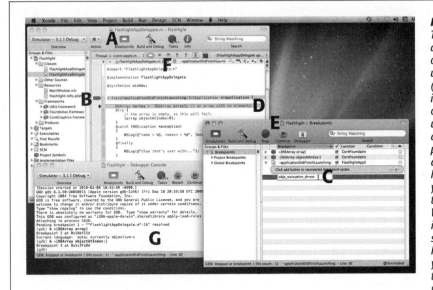

Figure 2-1:
The Xcode debugger in action. You can enable or disable breakpoints using the toolbar icon (A). To set a breakpoint in your source code, click in the left gutter until you see a blue arrow (B). The Breakpoints window shows all the breakpoints that have been defined, and you can add new ones to the end of the list (C). When a breakpoint is reached, the line of source code will be highlighted (D), and you can easily navigate the stack back-trace with a pop-up menu (E). Controls are also available to step through code (F). More advanced operations can be performed in the debugger console (G).

- For a class method, the syntax is similar: Just use a plus sign (+) instead of a minus sign (–) before the first bracket. To stop at the +*array* method in *NSArray*, you use +*[NSArray array]*.

Once you've stopped at a breakpoint, you can examine your variables by hovering over your source code with the mouse pointer. In many cases, you're going to want to use the *gdb* command line to examine the state of your instance variables and to send them messages. Here are a few important tricks:

- *po*, the *print object* command, displays the description for an object. In many cases, this will contain information that helps you figure out what's going on. Using the *myArray* variable from the code on page 52 as an example:

```
(gdb) po myArray
<NSCFArray 0x3b0e850>(
)
```

The print object command for an array normally shows all the elements in the array. In this case, none are shown, so you know that accessing the first one is going to be a problem.

- *p*, the *print* command, displays the value of an intrinsic type. Like the previous command, it's smart about the type of data you're examining: Structures and enumerations will be used as necessary.

- Methods can be called from both print commands. For example, if you've discovered a problem with an object not being retained correctly, you can fix the problem temporarily with:

  ```
  (gdb) po [myArray retain]
  <NSCFArray 0x3b0e850>(
  )
  ```

  The above example returns an object, so the *po* command is used. If a method returns an intrinsic type, you need to use the *p* command, and let the debugger know what type to expect. For example, you specify that an integer result is provided when you query the array for the number of entries:

  ```
  (gdb) p (int)[myArray count]
  $1 = 0
  ```

- When you see *address 0x0* in the console, it's a *nil* object (page 45). For example, if you ask the empty array for the last object, there is none, so you'll get back a *nil* result:

  ```
  (gdb) po [myArray lastObject]
  Cannot access memory at address 0x0
  ```

- The *p* command can also be used to assign a new value to a variable. To fix the exception caused by an empty array, you can assign a new one that has a single string in it:

  ```
  (gdb) p myArray=[NSArray arrayWithObject:@"MY CODE IS PERFECT"]
  $2 = (NSArray *) 0x3b0ee40
  (gdb) po myArray
  <NSCFArray 0x3b0ee40>(
  MY CODE IS PERFECT
  )
  ```

- The *b* command can be used to set a breakpoint from the *gdb* prompt. For example, to stop your application before retrieving an object at an array index, use:

  ```
  (gdb) b -[NSArray objectAtIndex:]
  Breakpoint 1 at 0xf135h90d
  ```

As you're probably aware, knowing your way around the debugger makes you a much more effective developer. To learn more, start by searching for the *Xcode Debugging Guide.* For more advanced information, take a look at *Debugging with GDB.* Both documents are available in the documentation viewer described on page 61.

Selector Projector

Because Objective-C is a dynamic language, it resolves methods at runtime, not at compile time. This setup provides a subtle but powerful mechanism: You can pass methods as parameters to other methods.

Selectors are special variables that identify which method implementation to use for an object instance. The Objective-C runtime uses internal data structures to match a message's selector to the code that gets executed.

Tip: If you've used other languages, you might want to think of a selector as a dynamic function pointer. The selector's name automatically points to the right function and adapts to whichever class it's used with.

This behavior lets you implement polymorphism in Objective-C. If you have three classes named *Circle,* *Triangle,* and *Square,* you can send a selector for a *draw* method to instances of each of these classes, resulting in different implementations for drawing the shape being executed.

Every selector is defined with a type of *SEL.* You can assign the selector variable with one of two methods. At compile time, you can use a compiler directive to create the value:

```
SEL mySelector = @selector(moreAwesomeThanEver);
```

At runtime, you can use a function that takes a single *NSString* parameter and looks up the selector:

```
SEL mySelector = NSSelectorFromString("moreAwesomeThanEver");
```

The *mySelector* variable now contains a shortcut for the *–moreAwesomeThanEver* message. And here's where it gets interesting: You can use that variable to execute the method. So instead of sending the message like you have been throughout this chapter:

```
[myObject moreAwesomeThanEver];
```

you can use the variable and achieve the same result:

```
[myObject performSelector:mySelector];
```

Now you could do something crazy like take user input, convert it into an *NSString,* and then call methods based on what the person typed. Luckily, the designers of Objective-C didn't implement selectors so you could be crazy. They did it so you could be lazy and could pass references to snippets of code that help the frameworks do the hard work.

Selector variables, and the ability to pass them around, play a central role in the Cocoa Touch frameworks that you'll learn about in the next chapter. Many of these features fall into the advanced category, but here are just a few of the cool things you can do with messages stored in variables:

- You can implement the delegation and target/action patterns described in the next chapter using selectors and the *–performSelector* method described on page 189.

- Arrays can use selectors to control sorting behavior. You can also use selectors when you want to send the same message to every item in the collection.

- If you want to keep track of when a view is animating, you can specify a callback method with a selector.

- You can configure timers with a selector that runs code after a delay or at repeated intervals.

- Objects register for system-wide notifications by specifying a method via a selector.

- You can use selectors to specify which method in an object initializes a new thread.

The number of colons in a selector's name is *really* important. While *@selector (moreAwesomeThanEver)* and *@selector(moreAwesomeThanEver:)* look similar, they have different method signatures. The first case is a method that takes no parameters, and the second case takes one parameter because it has a colon at the end.

If you've ever spent hours tracking down a stray semicolon in C code, finding a missing colon in a selector is the same sort of thing. Your eyes fool you into thinking that the code looks right, even when the runtime knows it's not.

Tip: When you see crashes in *objc_msgSend,* it's often the Objective-C runtime telling you that it can't find a method implementation. If you type your *@selector* incorrectly, that will be the cause. Make sure that the object has an implementation for the method being used.

Show Your *id*

Sometimes you need to refer to an object without knowing its class. It's convenient to reference an object just by its memory address without any type information.

You've already seen one of these anonymous pointers on page 50:

```
- (id)init {
```

Objective-C defines *id* as a pointer to an object data structure. (It's much like a *void* pointer in C, except that the pointer is only used to reference objects.)

So why does the *–init* method return this generic type? Because you have no guarantee what class of object the superclass will return to you:

```
if (self = [super init]) {
```

During initialization, objects are in a state of flux, so their class is purposely ignored and considered subject to change. Using a type of *id* expresses the object's lack of identity.

Even though *id* is nothing more than a pointer to a block of memory, you should never use a *void* pointer in its place. The compiler is able to perform additional checks and optimize code because the *id* lets it know that the variable points to an object.

The *id* type is also used as a way to avoid casting between classes. Some classes need to manage objects without knowing anything about the type of data. A good example of this is collections and other object containers. If objects in a collection were specified with a pointer to an *NSObject,* you'd have to cast to a subclass each time you retrieved an object from the collection. For lazy programmers, that's a lot of work.

Note: An *id* also solves the situation with overridden methods in a subclass that return a different type than the original class. By using a generic type, both the original class and any of its subclasses can share a common method signature and return the appropriate class.

Again, the *NSArray* class will be explored in the next chapter, but you already know that arrays are only useful if you can access the elements in the list. The method for doing this is:

```
- (id)objectAtIndex:(NSUInteger)index;
```

So, given an integer index, an *id* value is returned. The same is true when you add elements to the array: Combining these methods gives you a way to store any kind of object you'd like in the array.

Of course, your code usually knows what it put in the array, so it's common to do an implicit cast as you read the array element:

```
NSString *element = [myArray objectAtIndex:i];
```

This code lets the compiler check any code that accesses the element object.

Finally, there's one special *id* value that you've already seen mentioned several times in this chapter—*nil.* As with pointers in C, you'll occasionally want to represent the absence of an object. Using *nil* to represent a null object lets you do this.

Both *nil* in Objective-C and *NULL* in C are defined as *void* pointers with a value of zero. For example:

```
#define NULL ((void *)0)

#define __DARWIN_NULL ((void *)0)
#define nil __DARWIN_NULL
```

Semantically, however, you should use *nil* when dealing with pointers to objects and *NULL* for all other types of pointers. Technically, there's nothing wrong with this code:

```
NSString *myString = NULL; // PLEASE BE AWESOME AND USE nil INSTEAD!!!
```

It Really Is Just C

On page 30, you saw that Objective-C is just C with a little extra syntax and a runtime.

Now that you've learned a little bit about the language, you might want to see how Objective-C performs its magic. This is advanced stuff, but if you're the kind of developer who likes to look under the hood, you'll love this box.

First, start with some standard Objective-C code from an earlier example:

```
NSString *myString = @"typing power";

NSString *myResult = [myString awesomeString];

NSLog(@"myResult = %@", myResult);
```

You can do exactly the same thing with this code:

```
#import <objc/objc-runtime.h>

id myString = @"typing power";

SEL mySelector = @selector(awesomeString);

IMP myImp = class_getMethodImplementation
(object_getClass(myString), mySelector);

id myResult = myImp(myString, mySelector);

NSLog(@"myResult = %@", myResult);
```

What you've just done is replicate the work of *objc_msgSend:*, the code that's used by Objective-C to send messages to objects. It looks a lot like plain old C code, doesn't it?

The first step is to import the objc-runtime.h header file. That's where all the runtime definitions are stored.

You begin by creating a selector for the *–awesomeString* method. That's just a pointer to an internal structure:

```
typedef struct objc_selector *SEL;
```

That selector is then passed to the runtime function *class_getMethodImplementation()* that looks up a function using the object's class and the selector. All methods conform to this function definition:

```
typedef id (*IMP)(id, SEL, ...);
```

Once you have the function pointer to the class's implementation in the *myImp* variable, you can call it with the object and selector. This example doesn't have any additional parameters, but since *IMP* is defined with a variable argument list, they can be passed in as well.

When the compiler came across your method implementation:

```
- (NSString *)awesomeString;
```

it generated a C function that looks something like this:

```
id_i_NSString_AwesomeMethods_awesomeString
(id self, SEL _cmd);
```

The function name is mangled using the method type (instance or class), the class name, category, and method name. The function's arguments explain how the hidden *self* argument is passed. It also shows that there's another hidden argument called *_cmd* that your method implementation can use if it needs to know its selector.

And all this stuff happens automatically when you type those powerful square brackets.

It will, however, cause a nervous twitch and elevated blood pressure in longtime Objective-C programmers. Which is fine until you ask for their help in an online forum.

Note: Yes, there's irony in recommending a lowercase type definition instead of one that uses awesome uppercase letters like the rest of this chapter does.

Where to Go from Here

Now that you've gotten a taste of Objective-C, you may find that you want to know more about the language.

Sometimes, you do get something for nothing; *The Objective-C 2.0 Programming Language* book is a great example. Apple has some of the best technical writers in the business, and it shows in this book that describes the programming language. All of the language's features are described in detail along with pertinent examples. The book's glossary and index are especially helpful as you encounter new terminology and concepts. There's a link to this online book on this book's Missing CD page at *www.missingmanuals.com/cds*. (Or just do a web search for *the objective-c 2.0 programming language*. You'll find a link to the book on Apple's site.)

Another great resource is right inside this book—Appendix A. Here you'll find additional resources for each of the topics covered in this book. Each is listed along with a URL and a short review. If there's an area where you feel you need more help, head for the appendix.

Developer Documentation

It may seem odd that you've gone through an entire chapter describing a programming language, and the topic of developer documentation hasn't been mentioned. It's because there's a lot of it and it's really good. To the point of it being a distraction, in fact.

Since there's no way this book can explain every line of code you'll encounter, now's a good time to introduce the system documentation shown in Figure 2-2.

While you're working in Xcode, you can call up this documentation at any time from the Help→Developer Documentation menu. Once the window appears, you can search for anything by typing in the box at upper right: class names (such as *NSString*), method names (such as *release*), or concepts (such as *override*).

Once your search results appear, you can drill down using various navigational elements.

Results list

The results for your search are displayed on the left side of the documentation window. You can control how much information is displayed in the API list by selecting how your search term will be used to match documentation pages. When Contains is selected, *NSString* will match *CFStringConvertEncodingToNSStringEncoding* because it's a part of the API name. If Prefix is selected, *NSString(UIStringDrawing)* will match because it begins with the search text. The Exact setting would only show the page for *NSString*.

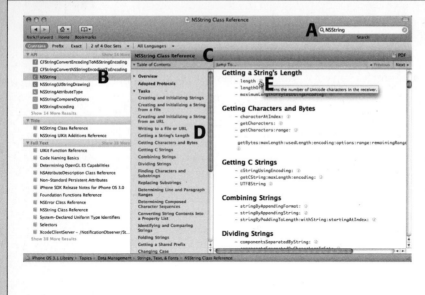

Figure 2-2:
The Xcode documentation viewer. You can search using the box at upper right (A). The results are listed on the left side of the window (B). Select one of the results, and the documentation page is displayed at right (C). From there, you can open the Table of Contents to quickly navigate through the page by clicking the links (D). For example, clicking the –length method name will take you to detailed documentation, or you can hover your mouse pointer over the info icon to get a quick description (E).

The results are broken down into three sections:

- **API.** This section shows symbol name matches. Each match is prefixed with a colored letter. A purple *C* tells you the symbol is a class, a blue *M* is a method, an orange *T* is a type definition, and so on.

- **Title.** Each of the entries under this section is a match for the document page's title.

- **Full Text.** The last section shows matches in the full text of the document page. These results can be handy to find what you're looking for in a larger context. If you're more interested in how you use a class, these results will show where it appears within a guide, release notes, or sample code.

Documentation page view

When you select an item from the results list, the page is displayed in the right side of the window. For some classes, this documentation page can be quite long, so here are some tricks to help navigate it quickly:

- **Table of Contents.** You can open the page's table of contents to quickly navigate through a page with many methods. And when you're first starting out, the "Tasks" section of the class documentation can be a helpful guide. For example, if you're looking for ways to create and initialize strings, there's a subsection for that under Tasks for *NSString*.

- **Companion Guides.** Many of the core classes in Cocoa have companion guides. These documents provide a high-level view of how to use the class, along with samples. If you're just starting out with *NSStrings,* clicking the "String Programming Guide for Cocoa" section will get you up to speed as quickly as possible.

- **Related Source Code.** Sometimes a picture is worth a thousand words; for a developer, sample code is the prettiest picture of all. If you're struggling with how to use *NSString* in a list of data, clicking *TableViewSuite* will show you how it's done in working source code.

- **Jump To...** If you're looking for a specific method in a class, this pop-up menu provides quick access to various sections and subsections on the page.

- **Use Find.** When you press ⌘-F, a new search field appears above the document page. Any text you type in this field shows up as a match in the documentation. You can click the arrow buttons (or ⌘-G and ⌘-Shift-G) to navigate through the matches.

Type less, learn more

The built-in documentation gives you easy ways to access it; don't miss these great tricks. Even using copy and paste, entering class and method names into the documentation search field gets tedious quick. The developers who created Xcode are just as lazy as you are, so they came up with some better ways to navigate through this sea of information.

Context menu

If you see a symbol in some source code that you don't understand, you can select the text and then Control-click to reveal a shortcut menu. This menu contains a lot of useful items, but toward the end you'll see two of particular interest. Clicking "Find Text in Documentation" brings up the documentation window with search results for the selected text. No typing required!

Another helpful item in that menu is "Jump to Definition". This command uses the selected text to look for the code where the symbol is defined. For a class like *NSString* where you don't have the source code, you'll be shown an *@interface* definition. If the class is one you've written, you'll be shown the *@implementation* definition.

Clicks ahoy!

But it gets even better. You can access the documentation by using just your keyboard and two clicks of the mouse button. Option–double-click a symbol name, and a small window appears that displays a short description of the class or method, a declaration that includes parameter types, and links to sample code and related documents. You'll also see the date when the API first became available.

> ***Tip:*** If you want to skip this preview window and jump right to the documentation window, Option-⌘-double-click the symbol name.

To find a definition, ⌘-double-click the name. If you get multiple matches, which is common with method names, you'll be shown a list of choices before the *@interface* or *@implementation* is shown.

If you're like most iPhone developers, you'll soon be using the mouse and these two keys without even thinking.

Learn to Be Lazy

There's that *L* word again. But the fact is, with Cocoa, a lot of great code has already been written for you. When you're starting out, it's natural to have the urge to write something yourself. Especially when it's a new language and you want to start exercising your new skills. When you feel that urge, take a moment to open the documentation and do a little research.

If you find something that's close to what you need, try to use a category to extend an existing class. If that won't work, try to create a subclass and add new functionality. When all else fails, implement your own class from scratch.

And with that advice it's now time to head to the next chapter, where you'll start to learn about some of this great code you now have at your disposal.

Cocoa Touch: Putting Objective-C to Work

Now that you've gotten a taste of Objective-C, it's time to put the language to work. If you think of Objective-C as the glue for building applications, you're now going to explore the building blocks—parts of the Cocoa Touch frameworks—that get pieced together with your new adhesive.

Before you start gluing the components of the Cocoa Touch frameworks together, you'll need to learn a bit about the architecture and design patterns recommended by Apple's engineers. Just as knowing how the correct building materials work together helps to build a house, a good working knowledge of the frameworks lets you build more robust applications and save time.

You'll also learn where to look for more information about Cocoa Touch. Plenty of resources on the Web and in your favorite bookstore can help you improve your iPhone development skills.

Get in Cocoa Touch

The Cocoa Touch frameworks are extensive, and hundreds of classes are at your disposal. Learning how to use this rich collection of components takes time. The best way to approach this Herculean task is to learn the common design patterns and functions implemented by the frameworks.

In the following sections, you'll see how Cocoa Touch uses models, views, and controllers (MVC) to implement your application. You'll also see how to use a target-action design pattern to hook the visual components to the code that does the actual work. In the process, you'll see how delegation is a powerful mechanism for leveraging code from the system frameworks. Of course, you'll also learn about managing data objects and collections, along with some of the more important design patterns.

What's in a Name?

You've probably heard the name "Cocoa Touch" used to describe the programming environment on the iPhone. But what is it?

Strictly speaking, Cocoa Touch is just two frameworks that provide the most important building blocks for your app:

- **Foundation.** This framework gives you the main building blocks. Within this framework, you'll find classes that manage data (such as *NSString, NSNumber,* and *NSDate*), that read and write information (*NSFileManager, NSUserDefaults*), that communicate with the network (*NSURLConnection*), and much more.

- **UI Kit.** Every application will have windows, views, buttons, and other interface elements. The *UIKit* framework provides the pieces that let users interact with your creation. You see these classes in the Interface Builder library palette.

Developers sometimes use the term "Cocoa Touch" as an all-encompassing name for other development technologies that work alongside base frameworks to create an app. You'll find yourself using many other supporting frameworks as you build your application. The most popular ones are:

- **Core Graphics.** A C-based API for drawing graphics (using the Quartz rendering engine). This low-level framework provides functions for drawing vector paths and bitmaps, 2-D coordinate transforms and masking, color and image management, and much more.

- **OpenGL ES.** Another C-based interface for accelerated rendering of 2-D and 3-D graphics. The implementation conforms to the OpenGL ES 1.1 and 2.0 specifications.

- **Core Animation.** This Objective-C API provides classes that allow sophisticated compositing and animation of a 2-D image hierarchy based on layers. This high-level framework greatly simplifies your code and improves the user experience. A nice side benefit is that it improves performance by leveraging low-level APIs such as OpenGL.

- **Core Data.** This sophisticated framework manages a graph of objects with transparent persistence in XML or SQLite files. Xcode provides tools to describe the objects and the relationships between them. The tools fetch objects using predicates and sort them using descriptors.

- **Core Audio.** A collection of frameworks that let audio be played, recorded, processed, and converted.

You can learn more about these building blocks by searching for their names in the documentation viewer.

The Big Three: Models, Views, Controllers

Cocoa Touch gives you hundreds of object classes to wrap your head around. Fortunately, most of these classes fall into three categories. And the objects in these categories interact in a simple and well-defined way.

Every iPhone application uses a simple Model-View-Controller design pattern. And since developers are just as lazy with writing as they are with coding, you'll often find this referred to as "MVC."

Note: If you've used other languages and development environments, you'll be very happy to know that MVC in Cocoa is no different than the ones you're used to. The objects change, but the concepts don't.

To see the simplicity of this design, take a look at Figure 3-1.

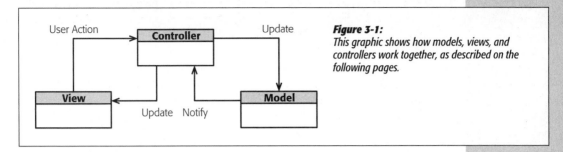

Figure 3-1:
This graphic shows how models, views, and controllers work together, as described on the following pages.

That's all there is to an iPhone app. And things get even better when you realize that views and controllers in the frameworks already do most of the hard work. Even the creation of model objects is simplified with the help of Cocoa Touch.

So what are the roles of each of these object types? Read on.

Views

You know all those buttons, scrolling lists, web browsers, and everything else that appears on your iPhone screen? Those are all *views.* Views know how to present your application's data. Some views also know how to react to user input. A *UIButton* view responds to a user's touch. A *UITextField* view takes input from the virtual keyboard.

In many applications, you'll create your own views for displaying specific data. For example, there's no standard widget for displaying stock graphs, so if you want to do that, you'll have to come up with your own solution.

Likewise, many designers and developers want to customize the look of their application. Whether your motivation is product branding or just wanting to stand out in a crowd, creating a unique look for your app involves building new views based on standard *UIView* and *UIControl* classes.

Creating views isn't as hard as it sounds. Using subclasses of existing views, you have to do little more than customized drawing. And if you're building composite views, which combine several views into a single view, your only work is managing the layout of the subordinate views.

Models

Models are your application's heart and soul because they're responsible for managing its data. Unlike views, models know nothing about the actions the user is performing or what they're seeing on the display. A model's only function is to manipulate and process the user's data within the application. Models often implement internal logic that provides these basic behaviors.

For example, whenever you use the built-in Contacts application, you're working with model objects that represent the people in your address book. If you update your application's settings, you're modifying another kind of model object. An application that downloads stock data from the Internet would use a model object to store the price history.

Some model objects work across multiple applications: The contacts and user defaults (settings) databases are good examples. Other models, such as those used by a stock application, are application specific.

Many model objects get stored permanently in files or databases. This setup is called *object persistence* and lets you recreate the state of your application across multiple launches. Even if you don't persist your objects, models can be very helpful when dealing with in-memory data structures.

When you're building your own models, you'll often use Cocoa Touch classes in the implementation. For example, classes like *NSArray* and *NSDictionary* let you store data in ordered lists or hash tables. You'll also find that classes like *NSURLConnection* and *NSData* help you retrieve and store data from a network.

Finally, you can use the Core Data framework to store and retrieve objects in an SQLite database.

Controllers

Controllers are a little more complicated. They act as an intermediary between the view objects and the model objects. When a value changes in a model, the controller is responsible for updating the view. Likewise, the controller knows when there's some user input and can update the model data accordingly.

Your controllers let you transfer information between models and views. In Figure 3-1, you can see that the controller is the only block that includes all the arrows (messages). Like a conductor in an orchestra, it's the one in charge.

To give you an idea of how this works, imagine an application that plots stock prices:

1. **The user clicks a button (view) to refresh the graph. This action is sent to the controller.**

2. **The controller, in turn, tells the model to load new stock data.**

3. **The model opens a network connection and begins downloading data.**

4. **After the stock data is loaded, the model notifies the controller that new data has arrived.**

5. **The controller passes the new data onto the view, and the user sees the results.**

Note: The models and views don't communicate directly because if they did, your application would get much more complicated as you add more objects. You'd have problems when multiple views share the data from one model, or a single view updates multiple models. Small changes in one class would ripple into others, classes would be increasingly dependent on each other, and you'd have a hard time reusing code. Developers refer to that as a *tightly coupled design,* and it's a good idea to avoid it.

Controller objects are often responsible for setup tasks. The models and views must be loaded and initialized at some point, and the central role of the controller makes it a natural candidate for this work.

Cocoa Touch gives you specialized view controller classes to handle this separation of responsibilities. As you develop your iPhone app, you'll find that much of the code ends up residing in these controller classes.

In particular, view controllers are used heavily in navigation. Some examples are:

- When you tap on a single row in a list and a new view slides into place, you've just created a new view controller. Likewise, when you tap the Back button, the current controller is discarded and replaced with the previous one.

- If you tap the button to compose a new mail message, a new modal view controller is created; its view slides up from the bottom of the screen and is displayed until you tap Cancel or Send.

- When you tap on the Info icon in the Weather app, a new view controller is flipped into place and lets you enter a new city. When you tap Done, that view is replaced by the main view.

The *UIViewController* class and several subclasses perform these basic functions. If you must master one class before any others, this one is it.

Value Objects

When you're building your own model, view, and controller classes, you'll often have to choose how to store data within those classes. Since Objective-C is built on top of C, all the primitive data types are available. The question is, then, whether to use integers, floating-point numbers, and character arrays or to use a class that wraps those same primitive values in an object.

Let's Get Primitive

There's no harm in using a primitive type in your object's instance variables. In fact, it can often make things much easier for your implementation, because you don't have to worry about retaining and releasing objects. For example, if you're just keeping track of a counter, you don't need to add the overhead of an object.

Tip: One of the most important counters in Cocoa, the *retainCount* in *NSObject,* is defined as an unsigned integer (*NSUInteger*). Although the keywords *NSInteger* and *NSUInteger* look like they could be class names, they're really just type definitions used throughout Cocoa Touch.

```
typedef int NSInteger;
typedef unsigned int NSUInteger;
```

Other examples of commonly used type definitions are *NSRange,* a structure that defines a range of data, and *NSTimeInterval,* a floating-point value that represents a period of time.

Instance variables that use primitive types are defined a little bit differently than instance variables of objects. Since there's no retain or copying involved, they're defined with the *assign* attribute for the *@property*. For example, if you had decided to use an *NSUInteger* instead of an *NSNumber* in the last chapter's *AwesomeString* class, you'd have an interface that looks like this:

```
@interface AwesomeString : NSString
{
    NSUInteger exclamationCount;
    NSString *originalString;
}
@property (nonatomic, assign) NSUInteger exclamationCount;
@property (nonatomic, copy) NSString *originalString;

- (NSString *)awesomeString;
@end
```

In the implementation the code in the *–init* method changes the initial assignment from a value object:

```
exclamationCount = [[NSNumber numberWithInt:8] retain];
```

You can simplify the code by assigning a primitive value:

```
exclamationCount = 8;
```

In the *–dealloc* method, you also don't need to release the *exclamationCount.* Using a primitive value has made your code much simpler and doesn't sacrifice any functionality. Until written languages start allowing you to put 8.5 exclamation points at the end of a sentence, this code will work great!

Objectified

Now that you've seen an example of value objects not being necessary, why would you ever want to use them? For two primary reasons:

- You can store values represented as objects in a collection. Collections make it easy to manage a group of objects. Methods are available to sort the objects, look up instances, and to filter data according to a pattern. If you're using primitive values, you'll have to implement this data management code yourself.

- The object's class provides a lot of functionality for manipulating the value. This gives you more flexibility and will save time in your implementation.

What are these value objects? Good question!

NSString

As you saw in the last chapter, this class is one of the fundamental classes in Cocoa. Many other classes use strings represented as objects. You may be tempted to use plain C strings in your implementation, and it's indeed possible to do so, but you're going to be swimming upstream. You'll end up doing a lot of unnecessary conversion as you use primitive strings with Cocoa.

Additionally, you'd be missing out on some important functions provided by the *NSString* class:

- Full Unicode support and conversion between string encodings (*–dataUsing-Encoding*).

- Reading text and its encoding from a file (*–stringWithContentsOfFile:used-Encoding:error:*).

- Splitting (*–componentsSeparatedByString:*) and joining (*–componentsJoinedBy-String:*) strings.

- Escaping strings so they can be put in a URL (*–stringByAddingPercentEscapes-UsingEncoding:* and *–stringByReplacingPercentEscapesUsingEncoding:*).

- Substring searches (*–rangeOfString:*) and getting the number of Unicode characters (*–length*).

- Converting strings into numbers (*–boolValue, –integerValue, –floatValue, –doubleValue*).

- Case conversions (*–capitalizedString, –lowercaseString,* and the best of all, *–uppercaseString*).

- String formatting (*–stringWithFormat:*) and localization (*NSLocalizedString*).

- String comparison using the user's current language settings (*–localized-Compare:*) and with many options, including ignoring diacritical marks (so "ö" is equivalent to "o") and sorting by numeric value (so "CHOCK9.TXT" appears in a list before "CHOCK9000.TXT").

NSNumber

You'll want to use this value object when you're doing value conversions. If you create a number object with a floating-point value, and then ask the object for an unsigned character value (*–unsignedCharValue*), an internal conversion will be performed without any loss of information in the original object.

If you're doing currency-based calculations in your application, you can use the *NSDecimalNumber* subclass for maximum accuracy. Since the subclass lets you use 38 significant digits and an exponent in the range of –128 to 127, you don't need to worry about rounding errors and other data loss inherent in floating-point calculations.

NSDate

A date object provides some basic functions for manipulating instants in time. You can compare which date is earlier (*–earlierDate:*) and how many seconds are between two dates (*–timeIntervalSinceDate:*). *NSDate* really shines when it's used alongside *NSCalendar*; if you've ever tried to compute the number of months and days between two instants in time while taking things like time zones and leap years into account, you'll appreciate this work being done for you.

NSData

As a value object for unstructured streams of bytes, *NSData* provides the mechanisms you need to manage a buffer of data. Data objects often need to be stored on disk, so methods to read (*+dataWithContentsOfFile:*) and write (*–writeToFile:atomically:*) are provided.

These chunks of data are often used to create instances of other objects. For example, an *NSString* can be created with an *NSData* object. Likewise, you can create an image using the *UIImage* object's *–initWithData:* method.

NSNull

This object serves only one purpose: to represent null values. Collections don't allow *nil* objects, so you can use *NSNull* when you need to put an empty value in an array, dictionary (hash table), or set.

NSValue

Sometimes you need to create objects that closely mirror a primitive type or data structure. This situation is most likely if you have some legacy or third-party data and need to add it to a collection.

Whenever this need arises, look at the *NSValue* class. It can wrap any variable type that's valid in Objective-C as a value object. Pretend you have a legacy system that defines a bizarre data structure called *Chockitude*:

```
typedef struct {
    unsigned char palmCount;
    int isFleshy;
    float height; // in inches
} Chockitude;
```

If you need to wrap this data in an object, it's simple:

```
Chockitude ch;
ch.palmCount = 2;
ch.isFleshy = true;
ch.height = 79.25;
NSValue *value = [NSValue valueWithBytes:&ch objCType:@encode(Chockitude)];
```

Whenever you need to retrieve the contents of the value object, you'd write this code:

```
Chockitude theChockitude;
[value getValue:&theChockitude];

NSLog(@"palmCount = %d, isFleshy = %s, height = %f", theChockitude.palmCount,
(theChockitude.isFleshy ? "YES OF COURSE" : "no"), theChockitude.height);

// this is displayed in the console log:
// palmCount = 2, isFleshy = YES OF COURSE, height = 79.250000
```

Another use for *NSValue* objects is with points and other geometry on the iPhone's display. If you wanted to keep track of a series of taps on the screen, you'd probably want to use a collection of *CGPoint* structures (that contain the x and y). You can use an *NSValue* category that takes these common structures and encodes them into objects: *+valueWithCGPoint:* and *+valueWithCGRect:* are used to wrap points and rectangles used by the graphics framework.

Collections

When you're dealing with objects, you're often dealing with many of them. It's essential to stay organized, and the easiest way to do this is by grouping similar objects into *collections*.

For example, when you look at a scrolling list on your iPhone, the data in each row is represented by an object. And if that row contains a person's name and age, you're likely to have other objects for those personal details.

NSArray is a natural collection for this kind of list view. You can specify the order within the collection so the first row in the list is the first object in the array. This array class also provides mechanisms for sorting and filtering the data, making it easy to model that behavior in a user interface.

When you want to collect objects by using a unique key, use the *NSDictionary* class. It implements an associative array that lets you look up objects (values) associated with a key.

Note: If you've used other languages, associative arrays are called a lot of different things: hash tables, hashes, maps, and containers are but a few. Regardless of name, they all let you find a value with a key.

You might use an *NSDictionary* to store the data for each row in your list. Since you can store *NSString, NSDate,* and *NSNumber* in a collection, it's easy to create a collection for that information:

```
NSDictionary *personalDetails = [NSDictionary dictionaryWithObjectsAndKeys:
    @"Craig", @"name", [NSNumber numberWithInt:50], @"age"];
```

You can then query the dictionary collection whenever you need one of the objects:

```
NSInteger age = [[personalDetails objectForKey:@"age"] integerValue];
BOOL crankyOldMan = NO;
if (age > 50) {
    crankyOldMan = YES;
}
```

You can use *NSSet* to collect distinct objects whose order isn't important. Typically, you'll use this type of collection as you would use a set in mathematics. Methods let you test for equality, intersection, and subsets.

As your application gets more complex, you'll find that you start to combine these collections. For example, your list view could use an *NSArray* for each instance of a row and an *NSDictionary* for the actual row data.

Likewise, it's possible to have an *NSDictionary* that contains an *NSArray* at a key. Imagine that your personal details include a *siblingNames* key; that data would be well represented by an array. If you don't have any brothers or sisters, the array is empty, and no matter how prolific your parents were, the array would grow to fit the size of your family.

Copying in Depth

When you make a copy of a collection by sending the *–copy* message, the result is a shallow copy of the object. The copied collection can be modified independently of the original collection, but the objects in both collections are shared.

To perform a deep copy of a collection, you use a *copy items* method. For example, with an *NSArray,* you use *–initWithArray:copyItems:* like this:

```
NSArray *original = [NSArray arrayWithObjects:@"Mimeoscope", @"Cyclostyle",
    nil];
NSArray *shallowCopy = [original copy];
NSArray *deepCopy = [[NSArray alloc] initWithArray:original copyItems:YES];
```

In *shallowCopy,* both objects in the array are the same instances that were in the original. The *deepCopy,* on the other hand, has new instances of *"Mimeoscope"* and *"Cyclostyle"* in the new array.

The *NSDictionary* and *NSSet* collections use the same pattern with the *–initWithDictionary:copyItems:* and *–initWithSet:copyItems:* methods.

Property Lists

When you go to the effort of organizing your objects in collections, it's likely that you'll want to save that work. Both *NSArray* and *NSDictionary* provide a *–writeToFile:atomically:* method. They also provide *+arrayWithContentsOfFile:* and *+dictionaryWithContentsOfFile:* methods to read the information back from the file.

What gets stored in this file is a *property list*, a standard format that's used throughout Cocoa Touch as a lightweight and portable mechanism for persistence. Property lists store instances of objects in a file with a .plist extension. The contents of the data can be stored as XML for maximum portability, or in a more efficient binary format.

If you opened your Flashlight application's resource folder, you might have noticed a file named Flashlight-Info.plist. This important property list contains information that describes your application to the iPhone OS. It's where you specify the icon that appears on the home screen, for example. Similarly, when a user updates your application's settings, the configuration is written into a property list.

When you're using property lists, it's important to use only the following classes: *NSArray, NSDictionary, NSString, NSData, NSDate,* and *NSNumber* (integer, floating-point, and Boolean values). Any hierarchy of these objects can be stored in the file. An array of dictionaries containing numbers, strings, and dates is completely valid.

Tip: If you need to archive an object graph that contains classes not supported in a property list, take a look at the *NSKeyedArchiver* class and *NSCoding* protocol.

Mutable Versus Immutable

While looking at *NSString,* you may have noticed that the content of the string is never modified directly. If you've checked out the *NSArray* class, you might be concerned that there doesn't appear to be any way to add or remove items to the array. Don't worry; the folks who designed Cocoa weren't *that* lazy!

Many of the classes for managing data are broken in two: one class to manage the data that never changes and a subclass for the part that can change. If the class name for the constant data is *NSString,* the class name *NSMutableString* is used for the one that can be modified.

Why use two classes when they could have just implemented one? For the same reason that constant values in other code are a good thing: You know that the data is never going to change as it's processed. You can make assumptions that you wouldn't have otherwise made.

Note: If you've ever had an array's size change as you're iterating over its elements, you know the pain of the bugs that these bad assumptions can incur.

When you're building your own classes, you'll often have to choose which of these two variants to use for your instance variables. A good rule of thumb is to use an immutable object like *NSString* or *NSNumber* when the value is replaced entirely. When you were making awesome strings, there would have been no point in using an *NSMutableString* since the *–setOriginalString:* method updated the whole string rather than making changes to individual characters.

Collections, on the other hand, are often updated incrementally. It's common to add or remove items from an array, so you'd want to use *NSMutableArray* and gain the *–insertObject:atIndex:* and *–removeObjectAtIndex:* methods.

Make It Mutable

What do you do when you have an immutable object and you need a mutable one? You'll find that a lot of methods in the foundation classes return immutable objects as results. These system classes assume that you're probably not going to want to modify the results. And for the most part, this is true. But as a developer, you know that it's the special cases that cause the most headaches!

Here's a simple example of a method that returns immutable data—the *NSString* method that splits a string using a separator:

```
- (NSArray *)componentsSeparatedByString:(NSString *)separator;
```

If you're parsing a comma-separated string to just loop through the values, an immutable copy works just great:

```
NSString *oneHand = @"Thumb,Index,Middle,Ring,Little";
NSArray *fingers = [oneHand componentsSeparatedByString:@","];
for (NSString *finger in fingers) {
    NSLog(@"The finger is %@", finger);
}
```

But what if you want to modify the fingers? This is one of those rare and important cases where you need a mutable copy. Everyone knows the last finger is called the "Pinkie".

Luckily, it's an easy fix using the *–mutableCopy* method that's defined by *NSObject*:

```
NSString *oneHand = @"Thumb,Index,Middle,Ring,Little";
NSArray *fingers = [oneHand componentsSeparatedByString:@","];
NSMutableArray *mutableFingers = [fingers mutableCopy];
[mutableFingers removeLastObject]; // OUCH THAT HURT!!!!
[mutableFingers addObject:@"Pinkie"]; // IT'S A MIRACLE OF MODERN SCIENCE
for (NSString *finger in mutableFingers) {
    NSLog(@"The finger is %@", finger);
}
```

It's important to note that you can't make a mutable copy of just any object. The class you're copying needs to support the *NSMutableCopying* protocol. If the class definition doesn't include *<NSMutableCopying>*, an exception will be raised when you try to make the copy.

Note: *–mutableCopy* behaves like *–copy* when you're dealing with collections. A shallow copy is performed.

Luckily, all of the value objects and collections support the protocol. And when a class doesn't support the protocol, it's probably for good reason. What would it mean to have a mutable copy of a *UIView* or an *NSURLConnection*? You'd have too many dependencies for such an abstraction to be understandable.

Protect Your Data

It's common to use mutable data in a method implementation. You have no idea how *–componentsSeparatedByString:* is implemented, but it's likely that the array was built up as the parser moved from the beginning to the end of the string. So why return it as an immutable array?

Cocoa Touch is preventing you from shooting yourself in the foot. Besides the fact that you probably don't need the mutable copy, when you have two objects with a copy of the same data, someone is going to get hurt if one object changes the data without the knowledge of the other object.

Imagine an array that's shared between two objects that are adding and removing items. The first object computes the length of an array. Then another object removes the last object in the array. When the first object goes to access what it thinks is the last item, an exception occurs because the array index is out of bounds. It's a hideous bug to track down, because each object appears to be doing the right thing. It's far from obvious that the root of the problem is a shared collection.

To return an immutable copy of a mutable instance, you rely on the fact that the mutable class is a subclass and that its superclass implements the *<NSCopying>* protocol. When you send a *–copy* message, you get back a nonmodifiable copy.

Suppose the *mutableFingers* array you created above is an instance variable in a *MyHand* class. You've decided to keep the variable mutable so you can accommodate those ignorant people who insist on calling the last finger "Little." When another object asks the instance for the finger names, you'd return them like this:

```
@implementation MyHand

...

- (NSArray *)fingers {
    NSArray *result = (NSArray *)[[mutableFingers copy] autorelease];
    return result;
}
```

The resulting object won't break your *mutableFingers*. Even if someone wants to call that last finger "Auricular" and stick it in his ear.

Delegation and Data Sources

Have you ever needed to perform a task and had no idea how to do it? Maybe it's fixing a dishwasher or knitting a pair of socks. You can either learn to do it yourself, or find someone to help you with the parts you don't understand.

Cocoa Touch has a design pattern called *delegation* that lets your application help out system classes that don't know what you want to do. The way it works is simple: You introduce one object to another object that's able to answer any questions. By assigning a *delegate,* you provide hooks into code that can respond to requests and state changes.

If your object is willing to be a delegate, it needs to follow a protocol. As you saw on page 36, the protocol is like a contract. Many of the methods in that contract will be optional, but others will be required.

This pattern is used extensively throughout Cocoa. If you type *delegate* into the documentation viewer, you'll get back over 100 results. You can tell which classes support delegation, because they have a *delegate* property. The actual protocols for delegation are usually suffixed with *Delegate.*

The best way to see what this looks like in practice is by checking out one of the classes that uses delegation: the *UIPickerView* control that selects values from a scrolling list (Figure 3-2).

Figure 3-2:
The standard picker view for Cocoa Touch. The values that are displayed come from a delegate and a data source.

When you look at the documentation for the *UIPickerView* class, you'll see the delegate instance variable defined like so:

```
@property (nonatomic, assign) id<UIPickerViewDelegate> delegate;
```

That declaration tells you that the delegate can be any class (represented by *id*) as long as it supports the *UIPickerViewDelegate* protocol. When you open the documentation, you'll see that the protocol has three primary tasks for the delegate. You'll be asked to provide dimensions and content, as well as to respond to selections. All of these tasks are optional, but your picker is going to be pretty boring if you ignore all the questions.

Initially, you'll probably want to customize the list of names in each row. The picker view doesn't know how to label the rows, so it will ask you what you want with the *–pickerView:titleForRow:forComponent:* message. As a delegate, you could implement a method like this:

```
- (NSString *)pickerView:(UIPickerView *)pickerView titleForRow:(NSInteger)row
forComponent:(NSInteger)component {
        return [NSString stringWithFormat:@"Row %d", row];
}
```

And each row in the picker would be labeled "Row 0", "Row 1", "Row 2", and so on.

You'd also like to know when someone taps on a row. If you don't implement the following method, you'll never know when the picker value changes:

```
- (void)pickerView:(UIPickerView *)pickerView didSelectRow:(NSInteger)row
inComponent:(NSInteger)component {
    NSLog(@"You picked row %d", row);
}
```

The best place to implement both of these methods is in a *UIViewController* subclass. This controller follows the MVC design pattern described on page 66, so it's easy to coordinate the activities between the picker view and a model that supplies the titles and that updates its state when a selection changes.

The picker view knows how many rows to display because the delegation pattern has a slight variation—the data source. As with a delegate, an instance variable with the name of *dataSource* must implement the *UIPickerViewDataSource* protocol:

```
@property (nonatomic, assign) id<UIPickerViewDataSource> dataSource;
```

And the *dataSource* is required to implement two methods: *–numberOfComponentsInPickerView:* and *–pickerView:numberOfRowsInComponent:*.

The first method tells the picker view how many columns to display, while the second method specifies the number of rows. For one column and 10 rows, you'd use these implementations:

```
- (NSInteger)numberOfComponentsInPickerView:(UIPickerView *)pickerView {
    return 1;
}
```

```
- (NSInteger)pickerView:(UIPickerView *)pickerView
numberOfRowsInComponent:(NSInteger)component {
    return 10;
}
```

As with the delegate methods, it's likely that you'd use your controller's model to fill in these values rather than using hard coded results.

You use this same delegate and data source pattern when you're dealing with lists of scrolling data, like the bookmarks in Safari or the inbox in Mail. Both use a *UITableView,* which is controlled using the same techniques as the *UIPickerView.*

Don't think that this delegate design pattern is limited to views in your user interface. Delegates are frequently used when downloading data from a URL. The *NSURLConnectionDelegate* defines a protocol to provide any connection authentication and to notify you as the download proceeds.

Targets and Actions

You use a mechanism similar to delegation when you're working with controls in Cocoa Touch. When you tap on a button, drag a slider to adjust the volume, or toggle a switch, these user interface elements need to notify other parts of the application of this state change. So how do these controls let everything know what's going on?

The notification occurs using a *target-action* design pattern. In other words, you can set up each control with a *target*—an object that will be notified of the change. As with a delegate, you can choose any object.

Unlike with a delegate, the action can be any method defined by the object. The important thing is that the method conforms to one of these two signatures:

```
- (IBAction)actionOne {
}
- (IBAction)actionTwo:(id)sender {
}
```

Interface Builder uses *IBAction* to identify the actions in your code (explained in detail later). The second form takes a single parameter that contains the object that sent the action. The *sender* can be used while you're processing the action.

If you have a single action that's used by several instances of *UIButton,* you can use the *sender* to determine which button triggered the action. Another use for the sender is to query the control's state. For example, if you received an action from a *UISwitch* control, you'll want to know whether the switch is on or off:

```
- (IBAction)toggleSwitch:(id)sender {
    // sender is an instance of UISwitch, so use its -on method:
    if ([sender on]) {
```

```
        NSLog(@"AFFIRMATORY DIXIE CUP");
    }
    else {
        NSLog(@"NEGATORY MR CLEAN");
    }
}
```

This code also has the advantage of being easy to read. Especially if you understand CB slang from the 1970s.

Next up, you need to learn how these controls get hooked up with the target object and the action method. Actually, you can use two ways of defining targets and actions: one uses code and the other uses Interface Builder. You'll start with the code to get a deeper understanding of what's going on behind the scenes.

User Interface: The Hard Way

Every control in Cocoa Touch is a subclass of the *UIControl* class. This class defines the following method:

```
- (void)addTarget:(id)target action:(SEL)action
forControlEvents:(UIControlEvents)controlEvents
```

This method takes three parameters. The first is the target object that will be notified about the control event. Second, the action parameter, defines the message that will be sent to the target object. Third, *controlEvents,* lets you specify the types of events that will trigger the action.

The most common events are *UIControlEventValueChanged* and *UIControlEvent-TouchUpInside.* The first, *UIControlEventValueChanged,* is used for controls where a changed value is important. Examples here are dragging a slider (*UISlider*), typing in text fields (*UITextField*), and flipping switches (*UISwitch*).

UIControlEventTouchUpInside events are typically used with *UIButton* controls because you only want to know that the button was released with the user's finger inside the bounds of the button. It's important not to use the *UIControlEventTouch-Down,* even if you're a fan of the NFL. This event doesn't give users a chance to drag their finger off the button to cancel the button press.

Here's how you'd put this all together in your controller object:

```
#define LOUDER 11.0f

- (void)updateVolume:(id)sender {
    float volume = [(UISlider *)sender value];
    if (volume >= LOUDER) {
        NSLog(@"NIGEL TUFNEL WOULD BE PROUD");
    }
}
```

```
- (void)hitOrMiss {
    if ([mySlider value] > LOUDER) {
        NSLog(@"OF COURSE IT'S A HIT!!!");
    }
}

- (void)setupControls {
    [mySlider addTarget:self action:@selector(updateVolume:)
     forControlEvents:UIControlEventValueChanged];
    [myButton addTarget:self action:@selector(hitOrMiss)
     forControlEvents:UIControlEventTouchUpInside];
}
```

The –*updateVolume:* method is attached to the *mySlider* view reference in the controller object. When the slider moves and its value changes, the volume will be checked; if it goes past 11, a message appears in the console log.

The action signature without a sender parameter was used for the –*hitOrMiss* method. This method also shows how you can use other view states to act on the button press. If the slider is above 11, the button lets you know that you have a hit.

Note: Page 58 warned you about getting the number of colons in the selector statement correct. If you had typed *@selector(updateVolume)* or *@selector(hitOrMiss:)* in the code above, the colons won't match the methods you'd implemented. When you tested the controls, you'd be greeted with a crash and an "unrecognized selector sent to instance" message in your console log.

Along the way

You're now probably expecting to see that easy way of setting up action messages on target objects. But before you can dive into Interface Builder, you need to prepare your controller's code so that objects and actions are exposed. This small amount of work will save you loads of time as your project evolves. It's well worth the investment.

As you've seen in the MVC design pattern, one or more view objects are managed by a controller. So where do you define these views for your controller? If you said "in the *@interface*," you win a prize!

To get some hands-on experience, download *MissingCD_Xcode_Projects.zip* from the Missing CD page at *www.missingmanuals.com*.

In this exercise, your controller is named *HitMakerViewController,* and it has three views: a slider, a text view, and a button. You define the instance variables just as you did on page 70:

```
@interface HitMakerViewController : UIViewController {
    UISlider *mySlider;
    UIButton *myButton;
    UILabel *myLabel;
}
```

Also define properties for these instance variables so they're retained objects:

```
@property (nonatomic, retain) IBOutlet UISlider *mySlider;
@property (nonatomic, retain) IBOutlet UIButton *myButton;
@property (nonatomic, retain) IBOutlet UILabel *myLabel;
```

And here are the action method definitions:

```
- (IBAction)updateVolume:(id)sender;
- (IBAction)hitOrMiss;
```

By now, you've probably figured out that the first two letters in Objective-C identifiers are important. The *IB* is a hint that helps you solve the mystery. It stands for Interface Builder.

If you look up the definitions of *IBOutlet* and *IBAction,* you may be confused by what you find in UINibDeclarations.h:

```
#define IBOutlet
#define IBAction void
```

When you add *IBOutlet* and *IBAction* to your source code, you're not adding any additional functionality. As you see above, the definitions of these two identifiers are essentially NOPs. All you're doing is marking your code so Interface Builder has something to parse.

IBOutlet is used to mark objects that will be used in the graphical editor and in your source code. Similarly, *IBAction* is used to identify methods that are shared between the two editing environments.

The magical way

From your source code's point of view, the *mySlider, myButton,* and *myLabel* objects will magically appear when your application launches. This kind of sorcery makes your life a lot easier, but it's important to understand the trick so you can take advantage of the underlying sleight of hand.

It all begins with the application configuration file. HitMaker-Info.plist contains a Main NIB file base name (*NSMainNibFile*) value. This tells Cocoa Touch to load the named file at launch. When *MainWindow* is specified, all the objects in MainWindow.xib will be loaded into memory.

Note: The name "NIB" is short for "NeXT Interface Builder" and was used as the extension for files saved with older versions of the software. Many developers and most parts of the framework continue to use this term, although the files now use the .xib extension. This new format is based on XML, hence the first letter in the new name.

In some contexts, developers talk about NIB files; in others they use the term "XIB files." They're the same thing: a file that contains part of your user interface.

A part of the loading process is setting all the instance variables you've defined as being an *IBOutlet*. The NIB loading mechanism uses your accessors to set the instance variables, so *–setMySlider:, –setMyButton:,* and *–setMyLabel:* are all called using the object in memory.

The coolest part is that the objects that are loaded have all the settings you've made in the Interface Builder. If you've changed a view's background color, that change will be reflected in memory, and your instance variable has access to it.

After an object is read from the NIB/XIB file, it's sent an *–awakeFromNib* message. This gives you a chance to perform any additional setup on your objects. This point is important, because not every object property is editable using Interface Builder. Also, in some cases, you'll want to configure objects based on internal logic.

You'd be right in thinking that now would also be a good time to update your view outlets. However, if you send the *–setValue:* message to your *mySlider* instance variable in the *–awakeFromNib* implementation, nothing will happen. These views haven't been loaded yet, so the instance variables still have nil values. When you send a message to a nil object, the message is quietly discarded. This seems like a bug, but the engineers who designed Cocoa Touch did you a favor: views are loaded *lazily.*

Here's how it works: Views can use a substantial amount of memory, especially if you're using heavyweight objects like *UIWebView* that include a cache of images and other web content. On the iPhone, every byte counts, so view objects are loaded only when they need to appear on the display. An added benefit to this approach is that it takes less time for your app to load when it contains many views.

To find out when your view is loaded from the NIB/XIB file, override the *–viewDidLoad* method. The following code shows how you could initialize the position of your *mySlider* object:

```
- (void)viewDidLoad {
    mySlider.value = 11.0; // LOUDER
    [super viewDidLoad];
}
```

Cocoa Touch not only loads views lazily, but it can also remove them automatically when memory is running low. The framework knows which views are on the display and safely reclaims storage by removing any that aren't visible. When it does, it also sends your view controller a message letting you know about the low memory condition:

```
- (void)didReceiveMemoryWarning {
    [super didReceiveMemoryWarning];
}
```

If your view relies on a large cache of information or other data that can be reconstructed easily, then the *–didReceiveMemoryWarning* method is a great way to clear out those objects as well.

Of course, you'll want to know when your view is removed, so you can release your instance variables to save even more memory. If you don't override the *–viewDidUnload* method, your instance variables will continue to occupy space, but will be useless because they don't have a parent view. The implementation is simple:

```
- (void)viewDidUnload {
    self.mySlider = nil;
    self.myButton = nil;
    self.myLabel = nil;
}
```

This brings up a subtle point: The *–viewDidLoad* and *–viewDidUnload* methods can be called multiple times, so avoid any initialization that should be performed only once.

Tip: You don't have to use a NIB/XIB file created with Interface Builder in order to construct a view hierarchy. Overriding *–loadView* lets you build your own view hierarchy with code. It's harder to do it this way, but with some user interfaces it's a necessity.

If you're going to great lengths to adjust views loaded from a NIB/XIB file, typically when you're doing extensive animation, it's often easier to just override *–loadView* with code that allocates the view objects, initializes, and adds them to the main view with *–addSubview:*.

Finally, since the NIB loader retained your outlet objects after storing them in memory, you need to do one final thing in your view controller: Release the views when the view controller is deallocated:

```
- (void)dealloc {
    [mySlider release]; mySlider = nil;
    [myButton release]; myButton = nil;
    [myLabel release]; myLabel = nil;

    [super dealloc];
}
```

Now that you're an expert with the magic behind NIB/XIB files, you're ready to take a look at how they're created!

User Interface: The Easy Way

At first glance, this step-by-step guide may not look so easy. Don't worry, once you learn the work flow, creating objects and hooking them up with Interface Builder is a snap. Time to get started!

A head start

Download the HitMaker—Start project from the Missing CD page at *www.missingmanuals.com/cds*. This project contains all the code for the actions and targets you'll be using during the exercise. To begin, you'll get Xcode and Interface Builder running, and open the files for editing:

1. **Open the HitMaker.xcodeproj file in the HitMaker→Start project folder.**

   Xcode launches and shows you the project contents.

2. **Open the disclosure triangle to get a list of files.**

   XIB files are located in the Resources group.

3. **Double-click HitMakerViewController.xib to open the XIB file with Interface Builder.**

   A new icon with a drafting triangle appears in your Dock as the file is loaded.

After the document file opens, you see a list of objects that are contained in the file.

Tip: The document's name appears in the title bar and in the Window menu. It's common to have several NIB files open at once, so use these tools to find the correct document.

You also see a gray View window. It's a preview of the View object contained in the document. Meanwhile, in the document window, you see the word "View" with no disclosure triangle next to it (Figure 3-3). That indicates that the *UIView* object does not have any children, or subviews. You're about to change that.

Tip: If you accidentally close the View window, you can reopen it by double-clicking View in the document window.

Your first view

That gray window looks pretty boring, doesn't it? Time to start adding new views for the labels, slider, and button!

1. **Choose Tools→Library. If you're the kind of developer who loves the keyboard, use the ⌘-Shift-L shortcut.**

   The Library palette opens.

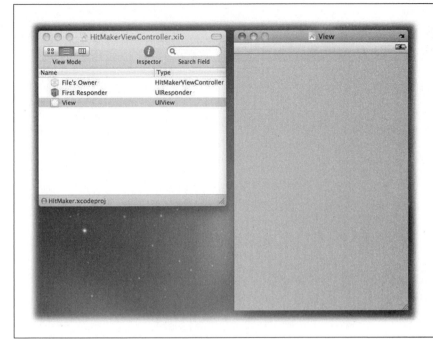

Figure 3-3:
Here's the blank palette you're going to work with. The document window is on the left, and the application's only view is on the right. Note that the View listed in the document window doesn't have any children (subviews) yet.

2. **Scroll the list of Objects in the Library, and select the cell with the text "Label".**

   It should be in the first row of "Cocoa Touch—Inputs & Values". When you select the cell, you'll see a short description of the control and its function at the bottom of the palette.

3. **Using your mouse, drag the Label cell onto the gray View window.**

   See Figure 3-4 for a demonstration. When you're dragging over the View window, a green plus sign appears, which indicates that the object will be added when you drop it.

4. **When your cursor is over the View window, let go of the mouse button.**

   The label drops onto the view, and four resize handles indicate that it's selected.

5. **Double-click Label and type** *VOLUME NOB.*

   Behind the scenes, you just replaced a line of code:

   ```
   [label setTitle:@"VOLUME NOB"];
   ```

6. **Drag the left and right resize handles until the dashed blue lines appear.**

   The edges of the label snap to the lines, helping you align your controls.

Tip: If you want to see the layout rectangles in View, press ⌘-L to display the bounds. If you press Option while the label is selected, you'll see the number of pixels between various interface elements.

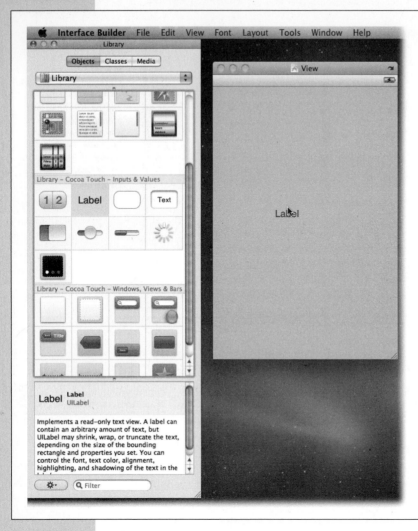

Figure 3-4:
On the left side, you see the Library palette with the Label control selected. You can drag any item in the Library to create a new instance of the object. Here, the Label is being dragged onto the View window.

7. **Make sure the label is still selected, and choose Tools→Attributes Inspector or press ⌘-1. In the Attributes Inspector's Layout section, click the Alignment (center).**

   You've just replaced another line of code:

   ```
   [label setTextAlignment:UITextAlignmentCenter]
   ```

8. **With the label selected, choose Font→Bold (⌘-B). If you want to go really crazy, press ⌘-I to make it italic, too.**

   Che bellezza! But even more beautiful is this code you didn't have to write:

   ```
   [label setFont:[UIFont boldSystemFontOfSize:17.0f];
   ```

As shown in Figure 3-5, the document window now shows a disclosure triangle for View. When you click it, you see the name is Label (VOLUME NOB) with a type of *UILabel.*

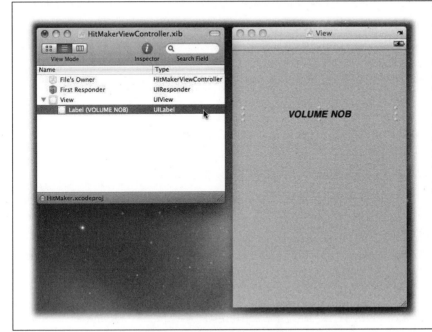

Figure 3-5:
Once you add a UILabel *containing the text* VOLUME NOB, *it appears under the main* UIView *for the window. When you double-click a view in the document window, it gets selected and displays resize handles—helpful when you're looking for a subview in a complex user interface.*

Congratulations! You just added your first subview to the hierarchy.

You're in control

You have your first object. But it's just a label, and you're not using it in your code. Time to remedy that with a control:

1. **Just below the Label in the Library palette, you'll see an icon that looks like a slider. Drag it onto the gray window.**

   A slider control appears, with resize handles on the left and right. Some controls can't adjust their width or height—the slider is one of them.

2. **Drag the slider so it's below the VOLUME NOB label.**

   The dashed blue lines help you center things.

3. **Drag the handles until they snap onto the dashed blue lines.**

You've created an object that you can use for the *mySlider* instance variable. All that's left to do is to hook it up to an outlet:

1. **In the document window, select File's Owner.**

   The type is *HitMakerViewController.* That's the same as the class you're writing!

2. **Choose Tools→Inspector and then select the second tab at the top (a blue circle with white arrow pointing right). Or just press ⌘-2 as a shortcut to select this Connections Inspector tab.**

   You see *myButton, myLabel,* and *mySlider.* Thanks to the *IBOutlet* identifiers you added to your source code, those show up automatically.

Tip: To prove this to yourself, open HitMakerViewController.h and remove the *IBOutlet* before *UILabel* *myLabel* and save the file. When you return, the outlet won't appear in the inspector for File's Owner. After you're done with this experiment, put the *IBOutlet* back; you'll need it in the following steps.

---

3. **To link the outlet to the view, click the circle to the right of *mySlider* and drag the mouse.**

   A blue line appears that lets you know you're connecting to an object.

4. **Continue dragging the mouse until you're hovering over the slider control. When a blue box appears around the slider, release the mouse button and you're done.**

   In your list of outlets for File's Owner, *mySlider* is connected to a Horizontal Slider (Figure 3-6). Yay!

Figure 3-6:
After you add the slider control as an outlet, the Connection Inspector for the HitMakerView-Controller *looks like this. As you can see,* mySlider *is connected to a Horizontal Slider. When you're working with Interface Builder, it's a good idea to use this inspector on File's Owner to check that all your outlets and actions are connected correctly. If a connection is missing, your controller code won't communicate with the view correctly.*

Tip: You can also connect outlets by Control-dragging from File's Owner to the control you want to connect. When you release the mouse button, you'll see a list of outlets to choose from.

Your document is starting to look like a user interface. You have two more outlets to connect:

1. **Drag another *UILabel* object from the Library onto the gray View window. Place it just below the slider, and adjust it like you did for the VOLUME NOB label, but don't bother changing the text of the label.**

   Unlike with the first label, you're going to update this one with code.

2. **Now go to the Connection Inspector after selecting File's Owner, and drag a blue line from the circle next to *myLabel* to the new label.**

   Finally, you need a button.

3. **In the Library Inspector, click the button icon (to the right of the label icon).**

   The description you should see at the bottom of the window is the Rounded Rect Button and the *UIButton* class name.

4. **Drag the Rounded Rect Button to the View window, and drop it in place just below the other controls.**

   As with a label, you can double-click the button to enter text for the button to display.

5. **Double-click the button and type *HIT OR MISS*. Then add another connection from the *HitMakerViewController* outlet named *myButton* to the button you just added. The drag-to-connect procedure is the same as you've just done for the other two outlets.**

6. **Press ⌘-S to save your work.**

   Interface Builder writes a new XIB file with the name HitMakerViewController. xib. Don't close the document window; you're not done with it yet.

   Next, you're going to head back to the connection attributes for the *HitMakerViewController* and connect *myButton* up to this new button.

Note: Some developers prefer to place and configure their controls all at once instead of switching back and forth between inspectors. Choose whichever work flow works best for you; just remember to connect your outlets after all the control placement is complete.

When you're just starting out, it's easy to forget these connections. Any code that references an unconnected outlet will be sending a message to a nil object and nothing will happen. Before you spend a lot of time debugging your code, make sure that everything is hooked up correctly!

Switch back over to Xcode so you can build the application with this new XIB file.

When you build and run the app using ⌘-Return in Xcode, you'll see that the app starts up and the controls get initialized (Figure 3-7). The *–viewDidLoad* code adjusted the slider to the middle position and changed the label text and color. But nothing happens when you drag the slider or press the button.

Figure 3-7:
After connecting the outlets to the slider and label controls, the HitMakerView-Controller *code in* –viewDidLoad *can update the position of the slider along with the label's text and color. When you launch the application now, nothing happens if you try clicking the button or slider.*

Get a Little Action

You've made a great start, but you need to get some action. No, not that kind. It's time to hook up an action to your new controls:

1. **Switch back to Interface Builder by clicking on its Dock icon.**

2. **In the HitMakerViewController.xib document window, select File's Owner.**

3. **Press ⌘-2 to bring up the Connection Inspector.**

Remember those *IBAction* identifiers you added to your controller's header file? You can see them listed under Received Actions. Make sure you can see both the gray view window and the document window where File's Owner is listed.

1. **Press Control and click the HIT OR MISS button. Drag the mouse, and you'll see a blue line. Since you want this action to be performed by the *HitMakerViewController*, move the blue line to that type in the document window.**

   When you let go of the mouse button, you'll see a menu appear with the events supported by that view controller (Figure 3-8). Click *hitOrMiss,* and you're done.

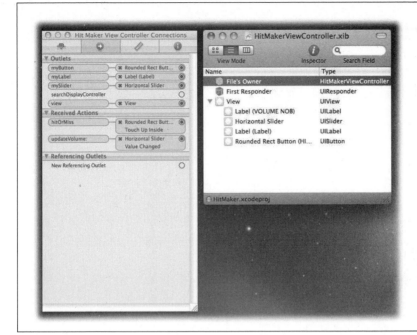

Figure 3-8:
The HitMakerView-
Controller fully connected.
When you select File's
Owner, you'll see that the
controller's three outlets
myButton, myLabel, and
mySlider *are connected
to views. Similarly, the
actions* –hitOrMiss *and*
–updateVolume: *will be
invoked by the button and
slider controls.*

2. Control-click the slider control. Drag the blue line onto *HitMakerView-
Controller,* release the mouse button, and then click *updateVolume:.*

3. Save the changes with ⌘-S.

Go to the document window, select File's Owner, and press ⌘-2 to open the Connection Inspector. You'll see *hitOrMiss* is connected to a Rounded Rect Button for
the Touch Up Inside event. Similarly, the Horizontal Slider's Value Changed event is
connected to the *updateVolume:* method (Figure 3-9).

Tip: You can even Control-drag between the items listed in the document window. Some developers
prefer this method since it doesn't involve opening the View window.

On page 81, you learned how events are processed while you were writing the
–*addTarget:action:forControlEvents:* code. In essence, you've just replaced the code
shown in –*setupControls* with a couple of clicks.

Now it's time to switch back to Xcode and test the app. Click its Dock icon and press
⌘-Return to build and run the app.

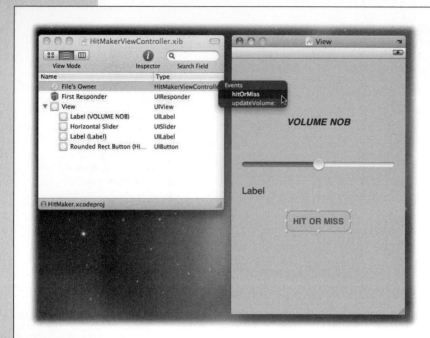

Figure 3-9:
After dragging a connection from the HIT OR MISS button to the HitMakerView-Controller, you'll select the hitOrMiss action. This move causes the named method to be called when the button is pressed.

Squash those bugs

Just because you're using visual development tools doesn't mean there won't be bugs. As you run the app, you notice a couple of problems:

- **Bug.** The slider value only goes from 0.0 to 1.0 (Figure 3-10). It has to go to at least 11.0 before it can rock.

- **Visual.** The text colors don't look so great on a gray background. It would also be nice if the text were center aligned. As is often the case with view attribute changes that are done in code, you don't know how things are going to look until runtime.

Luckily, both of these things are easy to fix with Interface Builder. Click the Dock icon to bring your XIB file back into view. In the following steps, you'll fix first the bug, then the visual problems:

1. **In the document window, double-click *UISlider*. Then press ⌘-1.**

    The Attribute Inspector opens.

Figure 3-10:
When you run the application, the slider and button controls now function, but some problems linger. The slider control value only goes up to 1.0, the contrast for the label is too low, and the text isn't aligned correctly. You can fix all of these bugs with Interface Builder.

2. Change Maximum value from 1.0 to 12.0.

   Because your app is just a bit more awesome than Mr. Tufnel's amp.

3. Double-click the second label listed under View in the document window. In the View section, click the Background color well. When the color picker appears, select black with 100% opacity.

   You'll also want to change the label text to white so you can see it against the new background.

4. Click the Text color well and picking the color as before (Figure 3-11).

5. Back under the label attributes, in the Layout section, click the center Alignment button.

6. In the View window, drag the label's height resize handle, and make it about 30 pixels tall.

Save your work, and then switch back to Xcode to build and run your application. As you run the app, everything looks good (Figure 3-12). And when you slide all the way to the right, you're ready to rock!

But that's not the best part: Think about how much code you haven't written. All of this is possible just because you typed a few *IBOutlet* and *IBAction* identifiers in your header file. That's awesomely lazy.

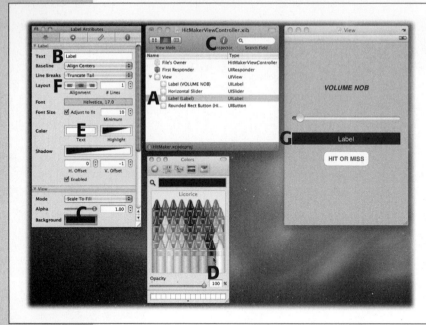

Figure 3-11:
Adjusting the colors and layout of the second label. Select the label (A) and then go to the Attributes Inspector (B). Click the Background color well (C), and select a black background (D). You can change the text color using its own color well (E). You adjust text alignment using the Layout control (F). The results are shown in the View window (G).

Figure 3-12:
After fixing the bugs with Interface Builder, the final build of your application rocks. Not only because you can move the slider all the way to the right and click the button, but also because you did all this work without writing a single line of code.

Notifications

In a complex system like Cocoa Touch, the state of the various components is constantly changing. Many of the things that happen are out of your control, but if your app adapts to the new conditions, it will provide a much nicer user experience.

Wouldn't it be great for your app to know when these things happen? Cocoa Touch provides a general mechanism for letting your application know of these changes via *notifications*. For example:

- The iPhone is locked or unlocked (*UIApplicationWillResignActiveNotification* and *UIApplicationDidBecomeActiveNotification*).

- The device orientation changes from portrait to landscape (*UIDeviceOrientationDidChangeNotification*).

- A user interface element causes the keyboard to appear onscreen (*UIKeyboardDidShowNotification*).

- The battery charge level changes or the device is plugged in (*UIDeviceBatteryLevelDidChangeNotification* and *UIDeviceBatteryStateDidChangeNotification*).

- A text editing view is updated (*UITextViewTextDidChangeNotification*).

- The pasteboard (Clipboard) changes (*UIPasteboardChangedNotification*).

Many classes generate notifications, and any object can generate them. The ones shown in this list are just a few posted by *UIApplication, UIWindow, UIDevice, UITextView,* and *UIPasteboard.* As you can probably guess by looking at the list, the standard convention is to use the class name at the beginning and the word "Notification" at the end.

When you want your app to be notified, you provide an object, a selector, and the name of the notification. A central service called *NSNotificationCenter* is responsible for moving the information from the objects that generate it to the ones that want to consume it. The objects that are distributed are instances of the *NSNotification* class. The notification center acts like a post office, and the notifications themselves are like letters.

When you want to receive a notification about a system event, you tell the notification center a little about yourself. Suppose you're interested in knowing when your application becomes active, either at launch or when the iPhone is unlocked. Typically, you'd add your view controller as an observer:

```
@implementation MyViewController
...
- (NSObject *)initWithNibName:(NSString *)nibNameOrNil bundle:(NSBundle *)
nibBundleOrNil {
    if (self = [super initWithNibName:nibNameOrNil bundle:nibBundleOrNil]) {
        ...
        [[NSNotificationCenter defaultCenter] addObserver:self
                selector:@selector(becomeActive:)
```

```
    name:UIApplicationDidBecomeActiveNotification
    object:nil];
  ...
}
```

The first parameter is *self*, which tells the notification center that the view controller will handle the notification. The selector *becomeActive* specifies the message that is sent to the view controller, and the name indicates the controller's interests. The *UIApplicationDidBecomeActiveNotification* is delivered to the controller when a user is able to interact with the user interface.

Note: The object parameter isn't used in this example, but it can be used to specify the sender of the notification. By specifying *nil,* you're saying that you want to receive every notification of this type, regardless of the originating object.

Since you've registered to receive the notification, you now need to implement the method that will handle it:

```
- (void)becomeActive:(NSNotification *)notification {
    NSLog(@"OPEN PUMPERNICKEL");
}
```

Now OPEN PUMPERNICKEL will appear in the console log each time you start or unlock your app (Figure 3-13). Great feature, huh?

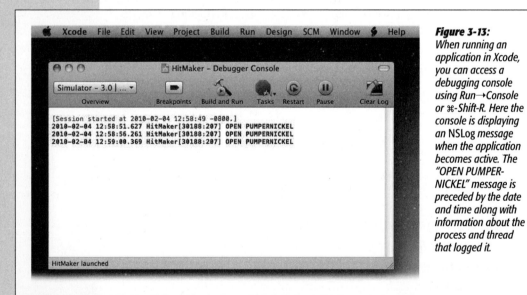

Figure 3-13:
When running an application in Xcode, you can access a debugging console using Run→Console or ⌘-Shift-R. Here the console is displaying an NSLog message when the application becomes active. The "OPEN PUMPER-NICKEL" message is preceded by the date and time along with information about the process and thread that logged it.

OK, so it's not a great feature. But what if that notification initiated a connection to refresh some data using the Internet? Or what about updating the device's location? It's a good trigger to let you know that the user is actively working with your application.

Similarly, you can use the *UIApplicationWillResignActiveNotification* to cancel any pending operations, especially ones that are keeping the wireless connections alive. If you stop the network operations, the radios can power down, and you'll increase the battery life. Everyone's happy!

It's very important to tell the notification center when your view controller object is no longer interested in knowing about these events. If you don't, your application can crash, because when you add an observer, notification is done by reference. If the object that was registered gets freed, the notification will be sent to an instance that no longer exists, and a memory fault will occur.

The fix is quite simple. Make sure you match every *–addObserver:selector:name:object:* with a corresponding *–removeObserver:name:object:*. You can also use *–removeObserver:* to remove all notifications at once. Since the observer was added during initialization, it makes sense to remove it during *–dealloc*:

```
@implementation MyViewController
...
- (void)dealloc {
    ...
    [[NSNotificationCenter defaultCenter] removeObserver:self];
    [super dealloc];
}
```

Note: These are notoriously hard bugs to track down. You'll see failures in *objc_msgSend* coming from code you didn't write, and they'll happen at times you don't expect. If you use notifications, make sure that they get cleaned up properly.

Notifications aren't limited to ones generated by the system. You can define your own to get the various parts of your app in sync. Imagine several controllers in your application need to know when a network connection finishes. *NSURLConnection* is based on delegation, so only one object will know when the operation completes.

That object could pass the information onto interested parties by defining its own notification name. A notification name is nothing more than a string object, so you'd define its existence in the class interface file like this:

```
extern NSString *MyConnectionDidFinishNotification;
```

In the object's implementation, you'd define the value for the notification name:

```
NSString *MyConnectionDidFinishNotification = @"MyConnectionDidFinish
Notification";
```

And when the connection completes, you would post the notification like this:

```
[[NSNotificationCenter defaultCenter] postNotificationName:
MyConnectionDidFinishNotification object:self];
```

Any object in your application that has called *–addObserver:selector:name:object:* with *MyConnectionDidFinishNotification* will then get a message and can update views, controllers, and models accordingly.

Singletons

The singleton design pattern is used in several key classes in Cocoa Touch. Singletons allow a class to return the same object instance anytime another object requests it. This is very handy when you *never* want more than one object of that type.

The *NSNotificationCenter* you just read about is an example. All notifications need to be routed through a centralized service for them to be effective. Just as importantly, objects need to be able to locate that central instance, and that's accomplished with the notification center's class method:

```
+ (id)defaultCenter;
```

Every object that sends a *+defaultCenter* message gets back the same object. That lets all objects add and remove observers in the same place.

Many times a class will use a singleton as a way to model that the thing only occurs once, either physically or virtually. You have only one application, set of settings, and hardware accelerometer, for example.

Unfortunately, there's no one standard for naming singletons. Older classes, such as those in the foundation, use the *default* prefix on the method name. Newer classes tend to use *shared* as a prefix. The most common singleton classes are:

- *NSFileManager (defaultManager)*
- *NSUserDefaults (standardUserDefaults)*
- *UIApplication (sharedApplication)*
- *UIAccelerometer (sharedAccelerometer)*
- *NSNotificationCenter (defaultCenter)*

Singletons as Globals

Many experienced developers recommend against using singletons. If you've ever programmed using global variables, you'll realize that singletons are really just a fancy way of representing a global state. You'll also appreciate that as applications grow and evolve, maintaining state that's shared by many components is a headache.

When you're implementing a singleton, you supply a class method that returns an instance of the class. So if you have a class named *MyFoot,* you'd define the method *+sharedFoot* like this:

```
@implementation MyFoot

+ (id)sharedFoot {
    static MyFoot *foot = nil;
    if (! foot) {
        foot = [[self alloc] init];
    }
    return foot;
}
```

The trick in this code is the static variable in the implementation; it holds a global variable that points to the shared object. When your application is launched, the value of that variable is *nil*. After the first message is received, the instance is allocated and returned for this and successive invocations.

Note: This singleton implementation isn't thread safe, nor does it prevent you from doing dumb things like releasing the global instance. If you're going to create singletons, take a moment to read Apple's recommendations in the "Creating a Singleton Instance" section of the *Cocoa Fundamentals Guide* (type *Singleton* in the documentation viewer).

Everything works great as long as your app needs to keep track only of one foot. But what happens when you add that new feature that uses both feet?

```
MyFoot *myLeftFoot = [MyFoot sharedFoot];
MyFoot *myRightFoot = [MyFoot sharedFoot];
[myLeftFoot shoot];
```

Not only are you shooting yourself in the foot, you're shooting them both at once. And you can't tell your left foot from your right because they're the same shared foot. The only way out of this kind of mess is by doing a lot of tedious refactoring. Singletons aren't inherently evil; just be sure you really need them before you go ahead with this design pattern.

Where to Go from Here

You've only scratched the surface of a set of frameworks that have been in active development since the 1980s. You've seen some of the most important parts of that framework, but you've certainly not seen anywhere near all of it.

As mentioned on page 61, the *Cocoa Fundamentals Guide* in the developer documentation is a great place to start learning more about this essential framework. Just search for its name in the documentation viewer. This book's appendix shows you books, websites, and other resources for learning more about Cocoa and Cocoa Touch.

As a new developer, make sure you also take advantage of the source code that Apple provides on the developer site. Studying these examples can teach you a lot about the right way to build an iPhone application. In a similar vein, the appendix contains links to many open source projects. Some very talented developers have contributed this code to the iPhone community, and you can learn from their work.

The Language of Design

Once you get your head wrapped around the language and frameworks, you're going to want to start putting these new skills to use. But before you do, you'll want to think about design.

Just as software has a system in place to organize the various components, the language of design can help you arrange your app's functional and visual aspects. The next chapter will explore this topic and show you how it can improve your entire development process.

Design Tools: Building a Better Flashlight

ow that you're up to speed with the language and frameworks used to create applications on the iPhone, you're ready to start coding, right? Wrong!

You may have years of experience developing applications for the desktop or Web, but the iPhone is a radical new platform running on a mobile device. Understanding how one or more fingers work with your user interface takes time. Assumptions based on how applications work in a controlled work environment change when you move into a chaotic mobile environment. These big differences also mean that the tools you'll use to design your app will change.

This chapter will introduce you to some of the unique elements of iPhone design. You'll also learn about a decision process that helps you come to terms with this new platform. And once you've gone through that process, you'll see how to start thinking about the code that will implement your ideas.

Plan Before You Code

You'll find a lot more to developing iPhone applications than just writing code. In this section, you'll learn what's different about designing iPhone software, how to create effective user experiences, how to start visualizing your ideas, and what tools will help during the design process.

Why Call in a Designer?

If you made it through the last two chapters unscathed, you're probably not someone who does design for a living. And that's just fine, because this section isn't about drawing things that are pretty. It's about making things that work pretty.

Many developers think a designer is someone who comes in at the end of a project and cleans up graphics—someone who can take all of those ugly images you created as placeholders and update them before you ship the product.

Designers can certainly do these things, but you're selling them short if that's all you ask them to do. Great designers will help you determine what the product should do by providing the perspective of someone who's not intimately involved with the technology.

As a developer, you tend to look at problems from the implementation outward. A designer thinks about the final product and works inward toward how it's constructed. When the logical left side of your brain encounters the designer's artistic right side, great things can happen.

Designers provide a point of view that you don't have. Don't be afraid to get this valuable input early in your product's lifecycle.

Design Goals

Here are some of the general things you and your designer should focus on:

- **Focus on the user and her task.** Your primary goal is to always think about use cases before forging ahead with a design solution. Some developers like to write user stories to formalize the kinds of tasks their customers will perform.

- **Follow established conventions.** Apple has done an excellent job developing actions, behaviors, gestures, and metaphors with the built-in applications. You'd be wise to follow their lead whenever possible.

- **Maintain consistency.** People intuitively learn the patterns in your application. Make sure that the look and feel for these patterns remains the same throughout your design. This is especially important when you need to deviate from the established conventions mentioned above.

- **Reduce complexity.** Designing a complex interface is easy. The most difficult job of a designer is distilling the elements down to their most basic level. Always question the need to add "just another control." Removing parts of your interface should give you more satisfaction than adding them.

- **Make the application's state clear.** If a background task like refresh is occurring, you need to convey that information to the user. When a user performs an action, visual feedback in the form of a highlight or an animation lets her know what's going on internally. Assume that people will make mistakes, so provide well-written feedback, and give them a way to undo the action.

- **A good design isn't static.** Don't be afraid to iterate! Your customers' needs will change over time. Leave room for growth—even when you have no idea how the product will evolve.

What's Unique about iPhone Design?

Even if you've had years of application design experience and the list in the previous section is second nature to you, you're still going to face many new challenges in your design. If you're wondering what's so different about designing for the iPhone, there's a simple answer: lots!

Let's get physical

The first thing you notice about the iPhone is how you interact with it physically. The contact is direct, not through an intermediate device like a mouse.

Input by human finger has some immediate implications that you need to consider in your design:

- **Finger size and reach.** A fingertip, like the one shown in Figure 4-1, is much larger than a mouse pointer. The length of the finger can also be a limiting factor when trying to reach different areas of the screen at the same time. Hands also come in a wide variety of sizes.

Figure 4-1:
This photo demonstrates one of the greatest differences between the iPhone and traditional desktop design. A mouse pointer represents one pixel on the screen; a finger uses about ¼" on the screen. That makes the hit target on the device about 40 times larger than on the desktop.

- **Weight distribution.** Control placement can affect the balance of the device in your hand.

- **Motion of the device.** Movement can act as an output such as vibration or as input with the accelerometer.

- **One or two hands in use.** Interaction patterns change dramatically depending on whether the user is using your application with one hand or two.

- **Left- or right-handed users.** Control placements for right-handed users may not be appropriate for left-handed users.

- **Display rotation.** The layout of your design can change when a user switches between portrait and landscape orientations.

Note that these issues are often interrelated. Designing for a left-hander using two hands in landscape can be just as important as for a right-hander using one hand in portrait orientation.

You also need to take into account environmental factors. A low-contrast design may look great in controlled lighting conditions, but when a user takes his phone out of his pocket on a bright and sunny day, he'll have trouble viewing the information you're presenting. Similarly, your design may be easy to read while stationary, but many people will use your app while jostled around in a bus or train.

It's important to take your creation outside the office and to see the effects of the real world on its usability. Do this early and often throughout the development process.

The Internet in your pants

The iPhone has two types of users. As you design your application, you need to keep the radically different needs of these two groups in mind.

Some people use the iPhone as a satellite device. These folks have a desktop computer in their home or office. The iPhone acts as a way for them to take their information on the go. Some aspects of app design for this group are:

- The Mail application is useful for notification and triage. Composition of long messages isn't comfortable.

- Capture and display of photos is easy, but extensive processing of the image is beyond the capabilities of the CPU.

- It's easy to add information to the Contacts app, especially with new people you meet. But the real value is syncing this information back to the desktop.

- Applications that don't sync important data back to the desktop are frustrating to use. Apple's built-in Notes application is an example.

The other group of iPhone users is the one that doesn't have another computer. For many people, their new smart phone is the first time they've had a personal computing device. These users don't expect the kind of features available on a real computer, but they'll still need flexibility while using your application.

For all kinds of users, the value of the iPhone isn't so much the function that it performs, but rather where it performs that function. Succinctly, it's the Internet in your pants.

Size matters

There's one other major area where the iPhone differs from the desktop: The resources available are much more limited. Table 4-1 compares the device sitting on your desk with the one in your pocket.

Table 4-1. *When it comes to resources, your desktop computer is much more powerful than your iPhone.*

Resource	Desktop	iPhone
CPU	Several GHz with multiple cores	Hundreds of MHz on one core
Memory	Several GB	Several hundred MB
Network	Wired connection with several Mbits of bandwidth	Wireless connection that can run as slow as dialup speeds
Display	Thousands of pixels in each dimension	Hundreds of pixels in each dimension

As you can see, in most cases the resources are an order of magnitude smaller than you're used to. Many of these limitations will come into play when you're implementing your design; others will have an immediate impact.

For example, the small display obviously limits the amount of information you can present on a single screen. Likewise, the limited amount of memory and processing power prevents you from putting thousands of items in your app's lists. Displaying hundreds of items is less taxing on the system *and* on the user who has to interpret the information. Less is more, both for the device hardware and the person using it.

Keep it simple

You have a lot of new things to come to terms with in your first iPhone application. Don't be afraid to avoid some of them initially.

Many successful designs on the iPhone started out simple. The developers focused on solving the problem at hand and kept the implementation as basic as possible. As their knowledge and experience about the platform grew, so did their product.

Your application will certainly evolve over time. Remember that it's much easier to add a feature in a future release than it is to remove one once it's in a customer's hand. Don't try to guess what people want from your application. Let them tell you and guide your education in this new and exciting platform!

The Design Process

Whether you're a designer or developer, you can use some simple steps and tools to get your application idea off the ground.

A paper prototype

Even if you don't have the skills to make nice-looking icons and images, you can still draw rough prototypes of your design. Your paper drawings may not be pretty, but they're just fine for getting the initial feel of the app.

This low-tech design technique is surprisingly effective. The key is that the piece of paper and the iPhone share something in common: You can interact with both by using your sense of touch.

You'll find yourself imagining what it's like to touch the controls drawn on the paper. It's also easy to move pieces of paper around as you envision the transitions between one screen and another. The fact that it's hard to write down lot of information in a constrained space naturally forces you to simplify ideas.

And unlike with code, if you find something that doesn't make sense, it's just a matter of pulling out the eraser and modifying the interface. Tools like Xcode and Interface Builder make your life much easier when you're implementing designs, but they can be a huge waste of time when you're in the discovery phase of development.

Paper also has another nice quality: It lets the designer and developer work together at the same time. It's difficult for a group to work on a Photoshop composition or on an Interface Builder document simultaneously. With paper, everyone can have a pencil and contribute. See Figure 4-2 for an example of a paper prototype.

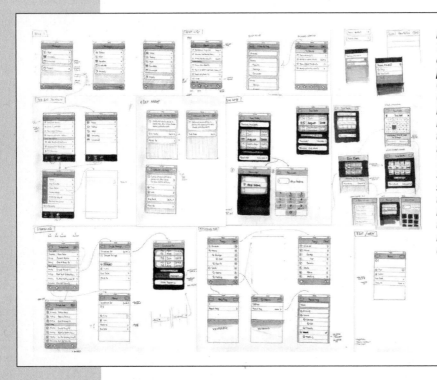

Figure 4-2:
This is a great example of paper prototyping done by Christian Krämer at Cultured Code. This illustration shows many of the screens they designed for their award-winning Things application. Some of these initial screens changed over the life of the project, but the basic flow of the application, represented by arrows, did not.

A pencil and paper and...

So what tools do you need to create these paper prototypes? Obviously, you need paper and a pencil, along with a couple of less obvious things. Take a look in Figure 4-3 at the tools that Cultured Code used to create the diagram in Figure 4-2.

Figure 4-3:
*The tools used to create
the paper prototype shown
in Figure 4-2. Everything
in this image is important,
especially the templates
for standard user interface
elements.*

Why the eraser? You shouldn't be afraid to make mistakes at this point in your design. Also, if you're working with a group, the eraser is an important tool to help refine someone else's idea. Highlighting pens are also helpful for visually breaking up the interface's components.

You'll also want to have an iPhone handy, with the latest version of the iPhone OS. You'll frequently want to refer to Apple's apps as examples of how to do things "the iPhone way." iPhone users are most familiar with the device's built-in software: If you mimic those designs and interactions, you'll be creating something that's easy for folks to learn.

The most important thing in this photo is the screen template with the iPhone's standard UI dimensions. You'll be drawing a lot of these boxes as you work through your design.

Christian's template was created by hand out of necessity. Luckily, entrepreneurs have realized that there's a strong demand for these types of products, and now several paper prototyping solutions are on the market:

- *http://uistencils.com*. This user interface stencil by Design Commission makes it a snap to draw the standard symbols and controls used on the iPhone. They also sell a sketchpad that works beautifully with the stencil. You spend your time thinking about the design rather than how to draw it. Both Figures 4-4 and 4-5 were created using this stencil and sketchpad.

- *http://appsketchbook.com*. Stephen Martin came up with this sketchbook after being asked to design several iPhone user interfaces. The result is a high-quality spiral-bound notebook with a wireframe template printed on each page. Perfect for the designer who wants to capture an idea.

Make it pretty

Once you have a feel for the overall design of the application, you'll want to start formalizing how it's going to look onscreen. This is the point where you start thinking about branding and adding unique elements to your app.

Note: For many companies, this is a really important part of the process. A firm like UPS doesn't want their app to be the standard shades of blue with black and white text. They're going to rely on their corporate color scheme of brown and gold. You can also bet their logo will be featured prominently.

This work is best left to a designer who's an expert with Photoshop and other illustration tools. If you're a developer, people will notice that you can't draw, and that distracts them from the awesome code that's behind the images. If you haven't gotten a designer involved during the paper prototyping stage, you'll need to spend time explaining your design. Which is another reason to get one involved early in the process.

Not all designs will go through this step. If you're going to use the standard UI controls and layouts, there won't be much for a designer to do.

As your designer is working on mockups, you'll want to make sure that they know that the standard UI widgets are available in several document formats. If your designer is working with Illustrator, they'll want to know about Mercury Interactive's vector UI elements (*http://www.mercuryintermedia.com/blog/index.php/2009/03/iphone-ui-vector-elements*). For designers who prefer to work in Photoshop, Teehan+Lax (*http://www.teehanlax.com/blog/2009/06/18/iphone-gui-psd-30/*) and Smashing Magazine (*http://www.smashingmagazine.com/2008/11/26/iphone-psd-vector-kit/*) both provide PSD files with elements arranged in layers. Another option is to use Patrick Crowley's stencil kit for OmniGraffle (*http://graffletopia.com/stencils/413*).

Tip: Working in Illustrator provides the most flexibility. Editing vector objects is generally easier than moving bitmap layers around. You'll also future-proof your work by producing it in a format that's readily scalable, and that's handy when a print publication asks for some screenshots that are larger than 320 × 480 pixels!

A first impression

You're also at a point where you can start thinking about the most important graphic in your application: the icon that appears on the home screen. This graphic is the first thing potential customers will see in iTunes, whether it's in search results or being listed as one of the top-selling apps. Think of it like a business card: You want to make a good impression, so let a professional handle this illustration!

The human brain processes shapes and colors much more quickly than words. As you work through the design of your application icon, focus on a distinctive color palette with clearly defined shapes. Avoid complicated and cluttered designs.

How does it feel?

Remember how your first app seemed huge when you ran it in the iPhone Simulator? The same effect occurs when you're working in Photoshop: The iPhone's 320 × 480 display looks bigger than it really is. If you're customizing user interface components, remember that your designer is working in an environment where the light is carefully controlled.

Both these factors tend to skew the design toward something that's appropriate for the desktop and not for a mobile device. That's why it's so important to expose your design to the real world.

The easiest way to get a feel for the Photoshop mockups is to save the 320 × 480 images and sync them to the built-in Photos app. Once you have the designs on the device and are carrying them around, it's easy to open the app and look at them in different conditions like bright sunlight and fluorescent lighting.

It's also possible to verify control sizes and placement by placing your finger over the simulated interface elements. Watch out for content that gets obscured by your hand as you move your finger around. It's also a good time to see how things feel for left-handed users and whether the design works well for one- or two-handed use.

POWER USERS' CLINIC

A Step Further

If you need to demonstrate your design to a larger audience, you'll want to package it in a way that makes it easily accessible outside of your project group. Luckily, a new open source project called Briefs (*http://giveabrief.com*) lets you do just that.

The Briefs application displays *scenes* with *actors*. The scenes are background images that you create using scans of your paper prototypes or flattened versions of the Photoshop mockups. Then, using an XML file, you define one

or more areas that respond to touch. These areas are called actors and let you transition between scenes. Once all this information is compiled into a *brief* and loaded on an iPhone, your paper prototype comes to life in your hand.

This obviously means more work to get it set up, but if you're working on an elevator pitch for financing your idea, or need to do a presentation to your company's CEO, a live demo can make or break your project. You'll also find this is a great way to do some early usability testing.

Living in Harmony with Your Designer

Designers and developers have decidedly different mindsets and ways of working. The last step of the design process for you, the developer, is thinking about how you're going to work with the designer to get the assets from their design tools into your development environment.

Finding a common ground

It's likely that your designer lives and breathes in Photoshop. Take a look at Figure 4-4. Designers know every nook and cranny of this madly complicated program and can do amazing stuff with it. Fortunately, you don't need to go out and buy a copy of Photoshop to use their work. All you need to do is ask the designer to supply your graphics in the PNG file format. This standard format is easy to export from a layered Photoshop document, and you can use it directly in your Xcode project.

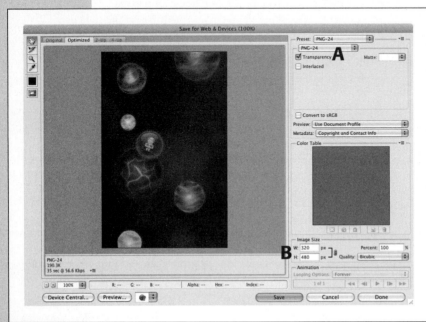

Figure 4-4:
A graphic from the Iconfactory game AstroNut being created with Photoshop. To open this dialog box, use the File→Save for Web & Devices menu. Make sure you're using PNG-24 (PNG file with 24-bit color) and Transparency (A). The screen size on the iPhone is 320 × 480 pixels, but you can set the image size to whatever is needed by the code (B).

If you need to edit the designer's work after delivery, it's usually just a matter of cropping or making other minor adjustments to the image. Many developers use a tool called Acorn (*http://flyingmeat.com/acorn/*) to do this simple cleanup.

When you're working with a designer's mockup, it's often necessary to measure various parts of the image to get the specified alignment right in Interface Builder or code. Is that button spacing 10 pixels or 11 pixels? What's the background color for this view? For these types of measurements, the Iconfactory's xScope tool is invaluable (*http://iconfactory.com/software/xscope*). This application provides a loupe, rulers, guides, and other tools that let you inspect and align screen graphics.

Give the designer control

For maximum efficiency, give designers direct access to your version control repository. A lot of tweaking tends to be involved with graphics built into a typical iPhone app. These constant changes can result in a lot of file transfers, test builds, emails, and other communication between the developer and designer.

You can save yourself a lot of time if you teach the designer how to build your app and how to replace images using the version control system. As you both feel more comfortable with the arrangement, you might find the designer making changes to your NIBs. And after you've adjusted a *UIColor* definition for the hundredth time, you might even suggest that the designer modify your source code!

Tip: Keep in mind that designers are typically visual thinkers. Their talents usually don't include working directly at the command line for extended periods. If you're going to provide the designer with access to your repository, you'll want to use a tool like Versions (*http://versionsapp.com/*) that provides a graphical interface.

Feedback: Don't Take Your Own Word for It

Now that you're done learning some of the basics of the design process, it's time to start thinking about what drives your decisions. The answer is both simple and complex. You're building this product for people to use. Their feedback should be the most important factor in the design.

The Providers of Feedback

You're probably thinking: "Dude, it's simple. Feedback comes from customers!" Which is absolutely true. But customers come in many different forms.

You

You're the most important customer. If you don't use and love your own product, how will your customers? The product came about because you had an itch that needed scratching: You tripped in the dark one time too many. So now you have a flashlight on the phone in your pocket.

At the same time, you're the worst customer. You're atypical because you know every single nuance of the product. People who are downloading your app from iTunes are not in that enviable position. But they'll be quick to notice things that you've overlooked!

Your marketing department

Developers tend to look at marketing people with an evil eye. They get a bad rap for asking for things that are hard to implement or that don't seem logical. They're looking at your product like a normal person would. Their motivation is to position, promote, and sell that product to other normal people.

The best marketing people are the ones who act as a conduit of information and ideas between you and the person who buys your creation. If you've done any marketing, you know that developers are more receptive to ideas if they're descriptive rather than prescriptive: Explain your problem, not your perceived solution. If you're a small developer, you won't have a marketing department. That's fine because you can easily find a surrogate.

Find someone you can bounce ideas off of. It should be someone who understands what the product does but not how it's implemented. This person can be your spouse, or best friend, or even the barista at your favorite coffee shop. All that matters is that this person is willing to have a long-term discussion about what you're doing and to provide unbiased feedback.

Be careful about using your designer for this task. Like you, they'll have an intimate knowledge of the product and may not be able to look at things with an untainted eye.

The people who plunked down their credit cards

Simply put, users are what make your product great. Whether they love your work, hate it, or are completely ambivalent about it, their feedback can help you improve your product.

User feedback comes in many forms:

- **Support email.** Users will contact you by email when they encounter a bug, don't understand how something works, or have a great feature idea. Trends in these communications are excellent indicators for how your product needs to evolve.

- **Beta testers.** These folks are the first ones to be exposed to new features and other changes in an application. If they complain about something not being right, chances are good that thousands of others will join them when the product goes into wide distribution.

Tip: Finding beta testers can be a challenge. Start by asking people you know if they're interested in testing. This can include people on a social networking site like Facebook or Twitter.

If you still need testers, try a site like iBetaTest (*http://ibetatest.com/*) that connects developers with enthusiastic users.

- **iTunes reviews.** Many one-star reviews about a feature are an indicator that you're doing something wrong.

Notice that all of these things should be taken in aggregate. Resist the urge to change your product direction on the basis of one email. It's an admirable desire to make every customer happy, but the reality is that you can't. Find the things that will please the most people.

Flashlight 2.0

Imagine that you released the Flashlight application you developed in the first chapter. You've been getting great feedback about the product. It's now time to develop version 2.0. What are you going to implement?

You've been getting a lot of support email about the need to control brightness. Many users have been complaining that the light is just too bright while they're navigating a dark bedroom: Spouses are being woken! It's clear from these reports that you need a quick way to lower the light level.

As you read your reviews in iTunes, you see a lot of people complaining about that wonderful shade of yellow you chose. Not everyone has the same great sense of color you do. Adding a color selector will make many customers happy.

Your marketing department (also called your spouse) spent some time evaluating the competition for your app: Many of those apps include a flasher. You think that's a great feature because you often exercise at night with your iPhone in an armband.

While at lunch with marketing, you start talking about the new flasher feature. You're merely matching the competition with the new flasher feature. Is there a way that you can give it a unique touch, maybe make it even more useful? It would be cool to have a "disco light" that randomly changes color while flashing. That's something that none of your competitors have, so marketing agrees that this feature should be investigated. But apart from making you a hit at parties, the feature isn't very useful.

And then it hits you both: What if the flasher could send an SOS signal? This international sign of distress would let rescue workers spot your customer's phone. Besides it being a great marketing angle, you're providing a feature that could potentially save someone's life.

So now you have your feature list for the new version:

- Brightness control
- Color selection
- Flasher
- Disco mode
- SOS signal

As you continue development, this list is your guiding light. And not just because you're developing a flashlight!

Bigger, Stronger, Faster

You've collected all the feedback, and feel good about the list of features that you want for version 2. Time to whip out the pencils and start designing the user interface!

The Light Side

The main view for the app is straightforward. The light source is predominant, and you'll need control mechanisms at the bottom of the screen for the various modes and settings access, as shown in Figure 4-5.

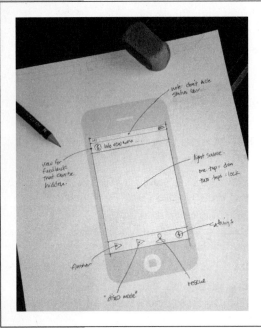

Figure 4-5:
A paper prototype for the main screen of your new flashlight application. The screen is broken into three areas: an information view, a light source, and a control area for switching modes.

The control area lets the user start the flasher, "disco mode," and the SOS rescue signal. The settings icon will flip the view, much like the built-in Weather and Stocks applications.

You're leaning toward using a toolbar to hold these controls at the bottom of the screen. A tab bar control is better suited to switching between views in an application—the app has only one view. The toolbar also has the advantage in that you can make it translucent, letting part of the underlying light source shine through.

Note: *UITabBar* is best used in applications that need to switch between views with different operational modes. Apple's use of the tab bar in the iPod and YouTube apps lets users pick different views of music and video collections. Users can also reconfigure tab bars to display preferred sets of information.

On the other hand, *UIToolbar* is appropriate for actions that work within a single context. A toolbar button lets the user act on the information in the view.

When designing your UI, think carefully about which of these control mechanisms is right for your application.

When switching modes, it's a good idea to give the user some feedback, so you decide to put an information view at the top of the screen. It would be nice to hide the information when possible, since it reduces the illumination produced by the light source.

So now you're feeling pretty good about the main view, but things are about to get more difficult with the settings interface. You have more controls to deal with, and they introduce some subtle interaction problems.

The Flip Side

Luckily, you've learned to use paper prototypes as a way to work through these issues. Sketching your screens, as shown in Figure 4-6, helps you identify trouble spots.

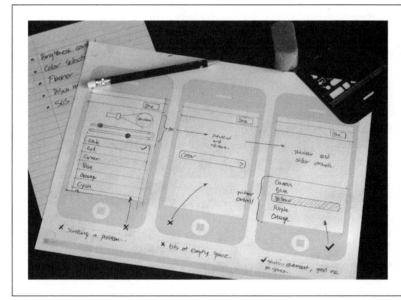

Figure 4-6:
A photograph of three proto-types for your settings screen. From left to right you see a grouped table view for color selection, a single button for choosing a color, and a picker control.

Your first decision is to include a small preview of the current settings. The reason for this preview is that without it, your users would constantly be flipping back and forth between the settings and the main view.

You've also decided that slider controls are necessary to control the brightness of the light and the speed for the flasher. It makes sense to put them under the preview since they're so closely related.

The first idea you have for picking the right color is a grouped table view. This makes sense, since it's similar to picking a ringtone in the sounds section of the Settings app. As you start drawing the interface, you quickly discover a problem: The UI gets cramped when you get more than four or five colors.

You'd like to give the user a selection of about eight colors to choose from. When you play around with the one you have on paper, it becomes obvious that the preview will scroll out of view when you move toward the bottom of the list.

So your next thought is to just have a single button that shows another list when you press it. But as you imagine the interaction on paper, this solution actually makes the problem worse; the preview gets hidden and there's a lot of wasted space.

As you think about other possible controls for selecting the color, you remember *UIPickerView*. It would be perfect: It has a fixed size and can display an arbitrary number of items. The height of the picker view also gives you more room for the preview and sliders.

A few minutes with paper saved you from writing a lot of code that would have been thrown away. And you now have two documents that you can use when discussing the interface with both your designer and marketing. Well done!

The Drawing Board

This flashlight is an admittedly simple application; the interface has only two screens. In more complex projects, you'll have many more. Go back and look at the number of screens in Christian's drawing at the beginning of the chapter, and you'll have a new appreciation for the complexity in an application like Things. In spite of these intricacies, a methodical approach to the problem lets you come up with a solution that fits the users' needs with consistency and without undo complexity.

It's important to keep in mind at this point that the design process is never complete. As you continue to talk to designers, marketing people, beta testers, and customers, the diagrams you just produced will change. Don't be afraid to pull out the eraser and improve your design. Iteration isn't just for coding!

Technical Design: Between Pictures and Code

But for now, everyone on the team is happy with the design, so it's time to start building it. Oh boy, you finally get to write some code! But not just yet.

The final step in the iPhone app design process is to translate the visual design into a technical design. You're going to be creating classes with methods as you work toward what you've put on paper. It's a good idea to think about how these pieces are going to fit together in the Model-View-Controller (MVC) pattern that was mentioned in the previous chapter.

If you neglect this important step, you'll find that you end up rewriting a lot of code. You don't need to write a 100-page specification that describes the technical needs in minute detail, but you do need to spend a little time planning where you're headed.

Start Naming

A good first step in this process is to start giving things names. You've probably already come up with some terminology as you've discussed the design with the other members of your team. It's time to start codifying the pieces of your application.

A *UIViewController* typically manages a screenful of data in an iPhone app. Since you're going to have two screens in your product, that's a natural place to start naming things. The view with the light source is the first one that you see after launch, so you decide to call it the "main screen." It would be fun to call the other view the "dimmer switch screen," but you think better of it and call it the "settings screen."

Now it's time to start thinking about the things you find on each screen and to give them names, too. This is when you start thinking of what you're going to call your models, views, and controllers. As with your visual design, strive for simplicity and consistency.

First light

So what kinds of classes are going to be required for the main screen? You'll need a view controller class to manage the contents of the screen. That controller will, in turn, manage view classes for the information at the top of the window, the light source, and the toolbar. Call these views the "info view," "light view," and "toolbar."

You'll also need a couple of models. One will manage and persist the current settings for the flashlight, so it makes sense to call it the "flashlight model." It also seems like a good idea to put the emergency signal generator into its own "SOS model."

Overall, the classes will fit together as shown in Figure 4-7.

This is also a good time to start thinking about the kinds of instance variables that each class is going to need. The view controller will have instance variables for each of the views and models (so that it can send messages to them). You'll also need actions to respond to the button presses on the toolbar: *–toggleFlasher, –toggleDisco, –startSOS,* and *–showSettings.*

You decide to have two types of messages in the info view: one for informational messages, the other for alerts. That *type* along with the *message* will be stored as instance data for the view.

The light view will have a *state*: The illumination state can be on, off, or flashing. You also want to be able to control its *brightness* and *color,* so you'll add properties for those to the view. A *delegate* instance variable will also let you report touch events back to the controller via the *–lightView:singleTapAtPoint:* and *–lightView:doubleTapAtPoint:* methods.

Since you're trying to simulate the physical behavior of a light bulb, it also makes sense to define how quickly the light reacts to changes. An *envelope* will let you change how quickly the light turns on (attack) and turns off (release).

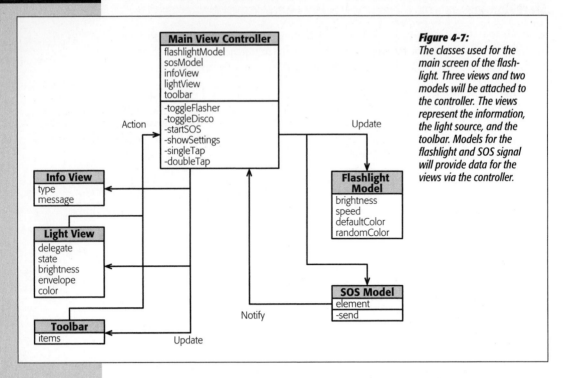

Figure 4-7:
The classes used for the main screen of the flashlight. Three views and two models will be attached to the controller. The views represent the information, the light source, and the toolbar. Models for the flashlight and SOS signal will provide data for the views via the controller.

The last view, the toolbar, will have *items* consisting of buttons that send actions and visual separators. For the flashlight model, you'll want to store and retrieve the *brightness, speed,* and *default color.* The disco mode will need a *random color,* so you'll add an attribute for that as well. The SOS model will need a *method* to start sending the signal and a way to get the current Morse code *element* (a dot, dash, or the gap between each).

The dimmer switch

The settings screen will also have a view controller that sits center stage. It's going to use the same flashlight model used by the main screen since you want the current flashlight data to be consistent across both views.

You'll also get to reuse the light view class from the main screen. It has properties like brightness and color that you can use for the preview. Objective-C and Cocoa Touch make it easy to write something once and use it over and over again.

The standard slider controls will be used for the light's brightness and flasher speed. Likewise, you'll be using the standard picker view to select the color. The sliders will invoke the *–brightnessChanged* and *–speedChanged* methods on the controller. The controller will act as a delegate for the picker view and implement the *–pickerView:didSelectRow:inComponent:* method to track changes in the control.

Like with the main view controller, there will be instance variables for each of the
views and the model. Figure 4-8 shows the overall architecture of the screen.

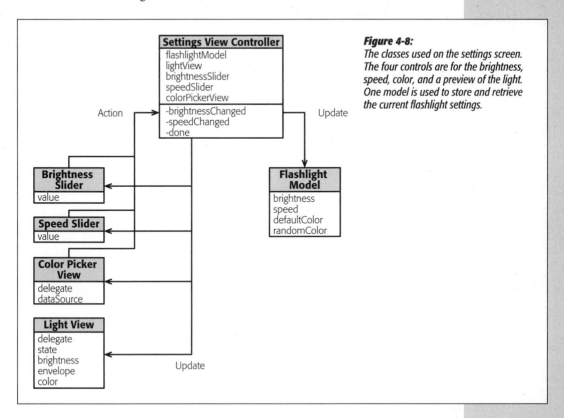

Figure 4-8:
*The classes used on the settings screen.
The four controls are for the brightness,
speed, color, and a preview of the light.
One model is used to store and retrieve
the current flashlight settings.*

Where to Go from Here

As you're developing your first iPhone user interface, you're sure to have additional
questions about the "right way" to do things. When that happens, the best advice is
to take a look at the *iPhone Human Interface Guidelines*. Every iPhone developer
should become familiar with this document from Apple that describes fundamental
interaction principles along with the views and controls that can implement them.
(You can find a link to it at *www.missingmanuals.com/cds*.)

You'll also find several books in this book's appendix that can help you maximize the
user experience of your iPhone application.

Ready to Code!

Don't treat your models, views, and controllers as if they're cast in stone. Details will
most certainly evolve as you implement the classes. Predicting the future has never
been an exact science.

The main purpose of going through this lengthy design process before touching a line of code is to filter out the bad ideas and refine the good ones. Any changes that pop up during the implementation will be minor. The major functions and features won't be affected, and you'll avoid the costs and headaches of mid-course corrections.

Attention to the details of design will also pay off when you start selling your app on iTunes. You'll find that it's easier to market your product because you have a clear vision of its purpose and the types of customers who will be interested in purchasing it. Customers will also appreciate the care you've put into the app's construction, and you'll get good reviews and word-of-mouth recommendations.

So now that you're armed with a design and a plan of attack, it's finally time to open Xcode and start implementing your ideas. Now is when the fun begins!

Part Two: Development in Depth

II

Getting Serious about Development

With your design documents in hand, you're ready to crack open a text editor and get to work! Chapter 1 gave you a quick introduction to the Xcode development environment. You're now going to dig into some of the details of your project file and see how to configure it for your own needs.

You'll also learn how to set up your work area so you can install your application on an iPhone. It's a complicated process—the configuration involves digital certificates and an Apple website that produces files that let your app run—but this chapter's step-by-step instructions will get you through it.

Throughout this chapter and the next, you'll see references to a project called Flashlight Pro. You'll want to have it handy as a reference. If you haven't downloaded it from the Missing CD page yet (*www.missingmanuals.com/cds*), now would be a good time.

Beyond the Template

Every project starts out with one of the templates that you saw on page 18. By nature, these templates are generic, but your project is unique. So you'll start by changing code and configuration files to suit your product's needs.

In this section, you'll become familiar with the basic settings in a working project and start to think about how to approach your new flashlight's development.

You're anxious to start coding, but you also know that there's always some setup required when you start using new software, especially true if it's your first project in Xcode!

Pick Your SDK

One of the first decisions you have to make is which version of the iPhone SDK you're going to code for. Different versions have distinct capabilities because of the underlying firmware in the iPhone's version of OS X.

Generally, it's a good idea to pick the oldest SDK version that has the features you need to implement your product. If you choose the latest and greatest, you won't be able to sell to people who haven't upgraded their OS in a while. Although the upgrade process is easy, many customers are slow to adopt new versions.

From a development point of view, the biggest changes to the iPhone APIs come when a major version update occurs. An example is when the 2.2.1 firmware was upgraded to version 3.0. The new version introduced hundreds of new features, including copy and paste, voice memos, MMS support, and many others. The 3.1 update brought only a handful of new features.

When Apple inevitably announces the 4.0 firmware, it'll undoubtedly include many great new features. When that happens, Cocoa Touch likely will provide new APIs (page 15) for you to use in your application. So even if, like most developers, you decide to target 3.0 now, don't be surprised when you want to follow your customers and change your target to the next major release.

Note: *API* stands for "application programming interface." Both Apple and developers use this as an all-encompassing term for the classes and methods available in the SDK.

With each new release of the iPhone software, an "API Diffs" document is published that lists additions and changes to classes and methods you can include in your app. You can use the documentation viewer to find and check the list.

Change Project Settings

After you've chosen which version of the iPhone firmware you want to write for, you need to configure your project file with that information. To do so, make sure your Xcode project is open, and then view the settings with Project→Edit Project Settings.

You see the Project Info window (Figure 5-1), which has the "Base SDK for All Configurations" setting at the bottom. From the pop-up list, choose iPhone Device 3.0 (or whatever firmware version you're working with).

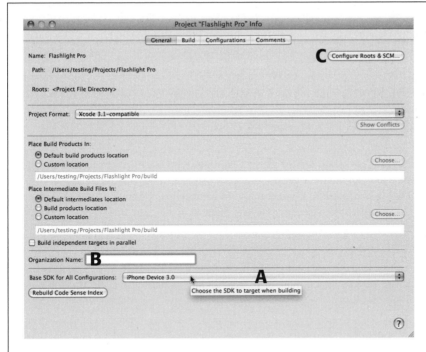

Figure 5-1:
Here's the Project Info window for Flashlight Pro. The most important setting here is the Base SDK (A), which specifies the minimum firmware requirement for your app. You can also configure your organization name (B) and source code management system (C) here.

For the most part, you won't need to alter any of the other settings in that window. You may want to change the Organization Name to your own company name. If you're using a source code management (SCM) system, you may also want to configure it here. The Appendix includes information on how to set up the Subversion SCM and Xcode.

Note: The project settings also have build information. You'll learn more about this important tab on page 132, after you've explored the target and its settings.

Change Target Settings

Xcode uses *targets* to define the set of instructions for building a final *product* using the files in the projects. Each target has its own name, steps used to create the application, and properties used to identify and then launch the application. When you create a new project, Xcode creates a default target with values for each of these items.

These default settings are fine for getting your project off the ground, but you'll eventually want to modify them. To do so, go to Xcode→Project→Edit Active Target.

What's in a name?

The Target Info window that appears has five tabs. The first tab, General, has a Name field, where you type the name of the product that you'll produce with the target—the name that appears on the iPhone's home screen.

Note: Many developers call the iPhone's home screen the *Springboard.* During the Jailbreak days, many people discovered that Springboard is the name of the hidden application that manages the launching of other applications. Even though you've never seen it, you've used Springboard thousands of times.

The first name you use for the project will rarely be your final product's name. For example, you might use a code name to avoid revealing what your app does until you're ready to release it. Or you may not know what the final product's name will be until late in the development cycle. Such was the case with the Flashlight Pro project: Marketing didn't decide on the name "Safety Light" until a few weeks before shipping.

Unlike most last-minute changes from marketing, this one is easy to deal with. Just change the target's Name field from Flashlight Pro to *Safety Light* and rebuild the app (Figure 5-2).

Identify yourself

Another change you should make is in the Properties tab. Each application on your iPhone has a unique identifier that the OS uses to manage it. In a new app, the Identifier is filled in with *com.yourcompany.${PRODUCT_NAME:rfc1034identifier}*. You need to edit this identifier to make it unique.

To make sure that your identifier is unlike any other, use the *reverse domain name* technique. In other words, if your domain name is *iconfactory.com,* then you use *com.iconfactory* for the first part of the identifier instead of *yourcompany.com.*

The second part of the identifier uses the Name you entered in the General tab by means of the PRODUCT_NAME environment variable.

As discussed on page 239, the product name is likely to change, and when it does, you end up with a duplicate application on your home screen. This ends up confusing you and everyone else on the project, so go ahead and pick a unique name now. Your customers will never see this name, so the only thing that's important is that you not use it for another product. The Flashlight Pro project uses *com.iconfactory. FlashlightPro.*

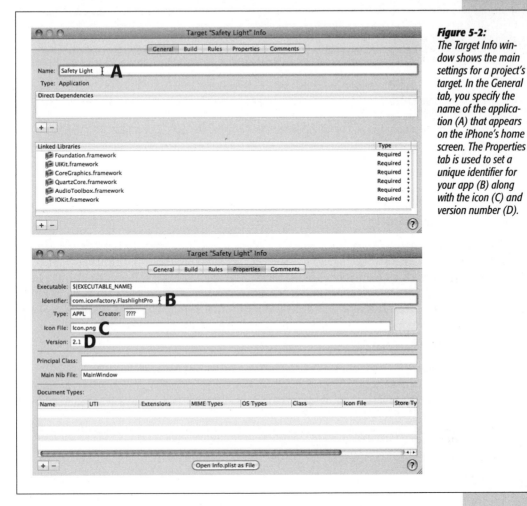

Figure 5-2:
*The Target Info win-
dow shows the main
settings for a project's
target. In the General
tab, you specify the
name of the applica-
tion (A) that appears
on the iPhone's home
screen. The Properties
tab is used to set a
unique identifier for
your app (B) along
with the icon (C) and
version number (D).*

Note: The identifier must conform to the same RFC 1034 specifications as domain names, so don't include any spaces or special characters. Just stick to letters, numbers, and dots.

Choose a distinctive icon

When you built the application in Chapter 1, you might have noticed a boring white icon appear on the home screen (Figure 1-12). That's what happens when you don't type a filename for Icon File in the target's Property tab (Figure 5-2).

Figure 5-3:
By default, your application icon gets a gloss and bevel effect that makes it feel at home on your iPhone. For some icons, such as the one used for Safety Light, these effects distract from the design. The image shows an icon on the left with the effect applied and on the right with it removed.

To replace boring with bling, supply the name of a 57 × 57-pixel PNG file that you want to use for the icon that appears on the scrolling pages of applications. You must store this file in your project's Resources folder, and Apple recommends that you name it *Icon.png*.

Although only one Icon.png file appears in the Property tab, Springboard actually looks for *two* files when you specify Icon.png. Since the 3.0 firmware added the Spotlight feature, your app's icon also needs a 29 × 29-pixel graphic called the Icon-Small.png file in your project Resources. If you don't supply a smaller image, the larger graphic will be scaled down when the user searches for your app. In many cases, this automatic scaling doesn't look as good as you'd want, so have your designer create graphics at both resolutions.

Note: Another graphic isn't specified in any configuration file: Default.png. This graphic is displayed as your application is launched, making it feel faster to the user. Some developers, especially ones producing games, use the image as a splash screen.

Make the file 320 × 480 pixels, and put it in your project Resources. If this file isn't included or has the wrong name, a black screen shows until your application is active.

Removing the Bevel effect. Many designers also prefer to add their own "shine" to the Icon.png and Icon-Small.png files. This presents a problem when you view the icon on the home screen: Springboard's default behavior is to add a gloss and bevel effect to the icon files (Figure 5-3). How do you get rid of that?

You've been editing Info.plist. All of the changes you've been making in the Properties panel of the target settings have been updating the Info.plist XML file. The file's full name, when you see it in the Resources group, is based on the name of the project. For the Flashlight Pro app, the filename is Flashlight_Pro-Info.plist. You can see it in the Resources group. When you build your target, the original file is processed by replacing environment variables like *${PRODUCT_NAME}* and by adding additional information like the minimum version of the OS that is required to run the application.

The resulting file is renamed Info.plist and placed in your application's executable folder. Every application includes this file, so Springboard has all the information it needs to manage, display, and launch your app. To remove Apple's Bevel effect, you'll dig in and edit the Info.plist manually, of course!

To edit your app's properties manually, follow these steps:

1. **At the bottom of the target's Properties panel, click "Open Info.plist as File".**

   An editor window opens that lets you edit the Info.plist values directly.

Tip: You can also open this editor in the project window by clicking the Flashlight_Pro-Info.plist file under the Resources group.

---

2. **Select the last line in the list, and a "+" button will appear at the end of the line.**

   A new property is added.

3. **Select "Icon already includes gloss and bevel effects" from the list.**

   You'll see an unchecked box as the value.

4. **Click the checkbox and save the file.**

   The next time you build and install the target (thereby giving Springboard a chance to read the new values), the gloss and bevel are removed.

You probably noticed that the list of keys available in the property list was quite long. Some of the keys, like "Status bar style", sort of make sense, while others like "Application uses Wi-Fi" don't. The one-line description doesn't help you understand how the configuration value is used. To learn more, you need to go a step deeper.

The data in Info.plist data is actually XML, and you can edit it like the rest of your source code. To do so, press Control and click the file in the project list. In the menu that appears, select Open As→Source Code File.

Tip: If you have a mouse with a right button, you can right-click the file and see the same menu.

Once you have Flashlight_Pro-Info.plist open, you can see that the key for "Status bar style" is *UIStatusBarStyle* and that its value is *UIStatusBarStyleBlackOpaque*:

```
<?xml version="1.0" encoding="UTF-8"?>
<!DOCTYPE plist PUBLIC "-//Apple//DTD PLIST 1.0//EN"
    "http://www.apple.com/DTDs/PropertyList-1.0.dtd">
<plist version="1.0">
<dict>
```

```
      ...
      <key>UIStatusBarStyle</key>
      <string>UIStatusBarStyleBlackOpaque</string>
      <key>UIPrerenderedIcon</key>
      <true/>
   </dict>
</plist>
```

A quick search of the documentation results in a document called *UIKit Keys* that explains this and other settings. This particular key configures the application to launch with a black status bar that matches the look of the rest of the application. You can also see that the checkbox for icon shine has the setting key of *UIPrerenderedIcon*.

While you're looking at other parts of this XML, you'll see a key named *CFBundle-Version* that contains the version number for your app. Even though this value isn't displayed by Springboard, you can display it within your own app (see page 204 in the next chapter).

Build Settings

As you explore both the Project and Target settings, you notice that both contain a Build tab. Your first question is, "What are these settings?" Simply put, they tell Xcode what you're going to create and how to do it.

As you scroll through the list of settings, the number of items that are listed may overwhelm you. Don't worry; you'll only need to change a handful of these default values. You'll see how to do so in detail on page 264 when you change the Code Signing values so you can put your compiled application on your iPhone.

Tip: You'll see a short description of each setting when you click an item. For example, if you click Product Name under Packaging, you'll see that this value is used as the base name of the product generated. A value in brackets shows the environment variable for the setting: *[PRODUCT_NAME]* is your clue that *${PRODUCT_NAME}* will be replaced with this value. The bracketed value also shows a compiler and linker flags if applicable.

Configurations

At the top of each build settings list are drop-down menus for Release and Debug. They let you define different settings depending on whether you're building a test version to debug or a release version to ship.

A great example of why you'd want to build a separate debug is when you're dealing with debug symbols and optimization levels. When you're debugging, the job is a lot easier when symbols that help you locate problems are in your executable. Similarly, optimized code is difficult to step through as you try to reproduce a problem.

Houston, We Have Ignition

Info.plist is obviously at the heart of the launch process performed by Springboard. If you're the type of developer who loves excruciating detail, here's the sequence of steps that brings the Flashlight Pro application to life:

1. A user taps on your application icon.

2. Springboard reads Info.plist and finds the compiled application by using the *CFBundleExecutable* property. This executable is launched just like any other process in Unix.

3. Your application contains a main function with the standard *argc* and *argv* list. A file named main.m in the Other Sources group shows this entry point that's called during the process launch.

4. When *main()* is called, an autorelease pool for managing memory is created, and the *UIApplicationMain()* function is called.

5. *UIApplicationMain* creates the single object instance of the *UIApplication* class. It also reads Info.plist and reads the *NSMainNibFile* property. For most applications, this property is *MainWindow* and specifies the NIB file to load.

6. When MainWindow.xib loads, the File's Owner is set to the instance of *UIApplication*. The other objects in the NIB, including the *Flashlight_ProAppDelegate* and *UIWindow,* are also loaded and connected.

7. The object graph that's loaded includes an outlet for the window in *Flashlight_ProAppDelegate.* The delegate for the instance of *UIApplication* in File's Owner also points to *Flashlight_ProAppDelegate.*

8. When the *UIApplication* finishes launching, it sends the message *–applicationDidFinishLaunching:* to the delegate. As the delegate, the method in *Flashlight_ProAppDelegate* is used.

9. The *–applicationDidFinishLaunching:* method creates an instance of *MainViewController* and assigns it to the delegate's instance variable. As the controller is instantiated, the *–initWithNibName:bundle:* method is called with the *MainView.xib* specified as a parameter. This loads and connects another object graph. The implementation of *–initWithNibName:bundle:* also sets up the models and configures notifications.

10. After the main controller is fully instantiated by the application delegate, the *–view* message is sent. This causes the view, which is loaded lazily, to be read from the NIB file. After that, the *–viewDidLoad* message is sent to the main controller.

11. The view created by the controller is then passed back to the application delegate. Using the *UIWindow* instance from the *MainWindow* NIB, this view is then added as a subview for the window. To make the window and its view visible, the *–makeKeyAndVisible* message is sent.

12. Since the main controller's view is now getting placed onscreen, the controller first calls the *–viewWillAppear:* method. After the view is displayed, the *–viewDidAppear:* method is called. Since the main view controller overrides the second method, this is when the user sees his preferred flashlight color displayed.

13. From this point on, user actions, and the events they produce, drive the application. You're in orbit.

Aren't you glad Cocoa Touch is awesome and that you didn't have to do all that by hand? It certainly makes launching a Space Shuttle look easy! You'll see a more humane approach to the launch sequence in the next chapter (page 157).

The opposite is true with your release version: You want to *remove* the symbols so the executable uses less space in the flash memory and so that it loads more quickly. You also want to take advantage of any speed and size optimizations that the compiler can make.

To see how these two configurations differ, open the Target build settings for Flashlight Pro using Project→Edit Active Target. Then click the Build tab, and set the Configuration menu to Debug.

In the Search box, type *optimization* and you see that the optimization level is set to None. If you change the configuration to Release, you see that the same setting is Fastest, Smallest. When you search for *strip debug* and flip between the two configurations, you see that symbols are only stripped from the Release version. That's exactly what you want.

Duplicity

Once you come to terms with the settings and configurations, you'll begin to wonder why two sets of them exist: one for the project and another for the target. Many developers get confused because both windows offer a long list of choices that are nearly identical. To decide which one to choose, you need to understand the hierarchy for the build settings.

At the bottom of the hierarchy is Xcode. It has built-in defaults for all of the settings in the list. For example, the optimization level for the compiler is set to Fastest, Smallest if you don't tell Xcode otherwise.

Above Xcode's defaults are the build settings for the project. Any settings you change here are reflected in all targets that are built with that project. Use the project settings for things that you want to happen in all targets. A good example of a setting at this level would be the optimization: You're going to want the Debug configurations of all targets to have no optimization instead of having Xcode's default.

The build settings for the target sit at the top of the hierarchy. If you change a setting in the target, it will override any setting in the project. Resist the urge to put settings here unless they're clearly tied to the product you're building.

When you start out with a project, you have only one target, so it's easy to ignore this hierarchy and just assign settings wherever you please. But you'll be adding more and more targets as your project evolves. For example, you'll probably add a new target when it's time to build a beta release, and another when it's time to upload the final version to the App Store.

If you have put a lot of settings in your targets, changing settings is a pain. Say that you have three targets and decide that you need to adjust the optimization level for your release builds. It's much easier to change that optimization level once in the project build settings than three times in each of the targets.

Note: As you're adjusting build settings, pay attention to ones that are in boldface. That's Xcode's way of highlighting the settings that are overriding a value in the child. For example, if you see "Optimization level" highlighted in the target build settings, you know that it's overriding the value in the project.

The Show menu is another good way to see all settings that are overrides. Select "Settings Defined at This Level" to get a list. (This list is filtered by the Search field's current value.)

Make It Official

Up until now, you've been able to run code only in the Simulator. The whole point of this book is developing apps for your iPhone, not a big honkin' facsimile running on your Mac. Why can't you see your app running in the palm of your hand?

If you're already a developer, you're used to working with source code editors, compilers, linkers, and debuggers. On the iPhone, you need to learn a completely new concept during your builds—*code signing.* Applications won't run on your iPhone until this step is complete.

This important new piece of your development toolbox is one of the more challenging aspects of iPhone development. Many experienced developers, even ones who have been working on the iPhone since its initial release, periodically get confused by how it works. When you encounter problems, take solace in the fact that you're not alone.

Join the iPhone Developer Program

As you saw in Chapter 1, Apple supplies the iPhone SDK for free, making it cheap and easy to get comfortable with this new development environment. But after you decide to distribute your product, you must join the iPhone Developer Program. It's going to cost you $99 per year to install applications on an iPhone or an iPod touch, either for testing or via the App Store.

This fee helps Apple run the program, but there's a more subtle reason for the charge—it's how the company verifies your identity. Apple takes security on the iPhone very seriously and doesn't want a rogue developer to damage the platform. When you use your credit card to enroll in the program, you're providing a name and address that have been verified by a financial institution.

Choose your program

If you've visited the iPhone Dev Center at *http://developer.apple.com/iphone*, you might have noticed the "Learn more" link in the right column. Click this link to start the signup process.

The first step is to choose one of the three account types that best suits your needs. You can choose from two types of Standard developer accounts: Individual and Company. The Company account lets multiple developers build and install apps. Both the individual and company accounts have access to the App Store.

Note: You'll find a slight difference in how iTunes lists the applications. An individual account shows your name as the "seller". Company accounts use your company's name.

You can also sign up with an Enterprise account. This type of account is only for large organizations with 500 or more employees. Applications created using this account can only be distributed in-house: You won't see them in iTunes.

Pay to play

Once you choose either a Company or Individual account, you'll go through steps where you supply your personal details, including credit card information. You'll also need to accept the program's licensing agreement.

If you're enrolling as an individual in the Standard program, make sure your enrollment and credit card billing information match exactly. If they don't, your application will be delayed because you'll need to provide additional identification. (Apple's trying to make sure you're not using someone else's credit card to pay for your enrollment.)

If you've selected a Company account, Apple verifies your business's identity as well. Apple asks you to provide various documents and contacts your company's legal representative to verify that you have the authority to accept the program's license agreement.

Once you've submitted your payment and proven that you are who you say you are, you get an activation code via email. After clicking the activation link, you receive another email from Apple welcoming you to the iPhone Developer Program. This is when it gets interesting!

Welcome to the club

After your successful enrollment, log into your new account, and you see some changes in the iPhone Dev Center (*http://developer.apple.com/iphone*). The navigation column at right shows links to new facilities:

- **iPhone Provisioning Portal.** You use this web application to manage the products you build with Xcode, the developers who create them, and the devices the apps run on.

- **iTunes Connect.** This is another web application that lets you control how your app is distributed in iTunes. It's where you upload your final application and provide the information used to sell it in the App Store.

- **Apple Developer Forums.** The developer forums provide an electronic meeting place for developers. You can search for previously answered questions, or ask your own. Members of the iPhone Developer Program also get early access to iPhone OS beta releases.

- **Developer Support Center.** A list of frequently asked questions covering a broad set of topics. If you have a question about the developer program, here's the first place to look.

Open the Door

The first place you're going to explore is the Provisioning Portal. It's the place where you get the stuff needed by Xcode to build and install your application.

What code signing means to you

Before you head off on the specifics of code signing, it's good to know why it's necessary and how Apple has implemented it.

Compilers and linkers have long been used to create application binaries. This code can be run on the target platform without any problems, but it has a couple of security-related deficiencies:

- **No way to verify who created the software.** With signed code, by contrast, Apple knows exactly who created software that's uploaded to the App Store.

- **No guarantee that the software hasn't been altered.** That's why code signing uses a cryptographic hash that lets the OS verify that the bits that are going to be executed are the same ones you built.

Code on the iPhone is signed using public key cryptography. That is, a combination of public and private keys (similar to those used for secure communications in your web browser) creates and validates a digital signature in the executable file. Your *private* key is used to create the digital signature, and a *public* key is used to compare it against the contents of the file.

The public and private keys used in the signing process are stored in the Mac OS X keychain as a certificate. A certificate authority at Apple Worldwide Developer Relations (WWDR) issues the certificate. WWDR also dictates when the certificate expires.

Apple also controls where the signed code can run by requiring development devices to have a *provisioning profile* installed. Each profile contains a list of 40-character hexadecimal numbers that uniquely identify the hardware that's allowed to run the specified application. Some provisioning profiles are specific to a single application, while others can specify multiple applications. The profile also contains a reference to the certificate of the developer who created the application.

When you launch an application for testing, the list of devices in the provisioning profile is checked, and execution proceeds normally if the current device is included. If there's no match, the application is terminated.

Note: Applications downloaded from the App Store don't require a provisioning profile: The FairPlay DRM used by iTunes controls their execution.

In short, you need two things before you can run an application on the iPhone itself: the ability to sign code using entries in Mac OS X's keychain and an installed provisioning profile that contains your device's ID.

Keychain setup

If you were adventurous and tried to do a device build of one of the previous projects, you likely saw a Code Sign error during the build. Unless you skipped ahead to this chapter, or have some previous iPhone development experience, the build won't complete because you haven't set up the keychain or installed the provisioning profile.

The first step is to get the Mac OS X keychain configured so that Xcode can use it during the code-signing build phase.

Note: The Mac OS X keychain is a security mechanism that's built into the operating system. After you enter a master password, usually at the time you log in, applications can get access to other passwords and to secure information. For example, when you enter a password for a website, it gets stored in the keychain so Safari can use it the next time you visit the site.

You manage the information stored in your keychain by using Keychain Access (it's in your Applications→Utilities folder). It's used to browse and manage the information stored in the keychain. For Xcode development, you'll be focusing on items in the Certificates category.

A new authority. The first thing you need to do is to set up Apple's Developer Relations as a certificate authority. Here are the steps:

1. **Log into the iPhone Dev Center at** *http://developer.apple.com/iphone.*

2. **In the navigation column at right, click iPhone Provisioning Portal.**

3. **In the left column, click Certificates.**

4. **At the bottom of the Certificates page, click the link that reads, "If you do not have the WWDR intermediate certificate installed, click here to download now."**

   A file named AppleWWDRCA.cer appears on your Mac.

5. **Double-click the file.**

   Keychain Access launches.

6. **If an Add Certificates dialog box appears, make sure the Keychain pop-up menu is set to "login" and then click the Add button.**

At this point you should see the new certificate installed in your login keychain (Figure 5-4).

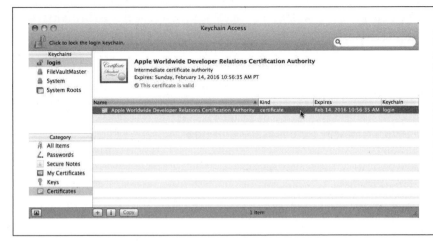

Figure 5-4:
Once you've down-loaded and installed the WWDR intermediate certificate file from the developer portal, you should see a certificate from WWDR in the Certificates category of the login keychain.

Request a certificate. The next step is to request a new certificate from the authority you just added. The process involves generating the request on your desktop and then uploading it to Apple via the Provisioning Portal:

1. **Select Keychain Access→Certificate Assistant→Request a Certificate From a Certificate Authority.**

    Keychain Access starts an assistant (wizard) that takes you through the request process (Figure 5-5).

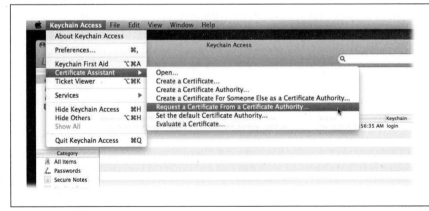

Figure 5-5:
Starting the assistant that generates a request for a new certificate from Apple.

2. **In the dialog box that appears, type your email address in the User Email Address field.**

    This email address should be the same one you used when signing up for the iPhone Developer Program.

3. **Fill in the Common Name field using the same name you used to enroll as an iPhone developer. Select "Saved to disk" so a file will be created in the next step.**

   Your form should resemble Figure 5-6.

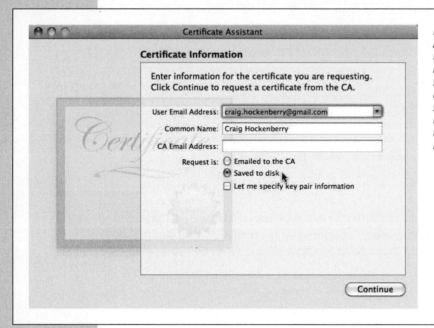

4. **Click Continue. When prompted to save the certificate request, change the name of the file to *iPhone.certSigningRequest,* and save it on the Desktop.**

   After the file is saved to disk, you'll see a confirmation. Click Done to close the assistant.

Note: The certificate that you're requesting will last only 1 year. You can save yourself some time by saving this request and resubmitting it each time the current certificate expires.

Certifiable. Now that you have the request file on your desktop, it's time to upload it with your web browser:

1. **On the Provisioning Portal page, click Certificates in the left column.**

   You'll see a warning that you don't have a valid certificate.

2. **To remedy that, click Request Certificate.**

3. **At the bottom of the page, click Choose File, and navigate to the *iPhone. certSigningRequest* on your desktop. After selecting the file, click Choose.**

   The dialog box closes.

4. **In the lower-right corner of the page, click Submit.**

   You should see a message that the "Certificate Request has been submitted for approval." The status of the certificate will also be "Pending Approval".

5. **You need to approve your own request, so click the Approve button in the Action column.**

   The status column now says "Pending Issuance".

Note: When you're using a company account, an admin will need to approve your certificate request. The admin user also has the ability to add new members to a Team.

---

6. **It normally takes less than a minute to issue the certificate, so refresh the page after a short wait, and you should see the status listed as Issued. Click the Download button in the Action column.**

   Your new certificate will be downloaded. You should see a *developer_identity.cer* file in your Downloads folder.

7. **Double-click this file and you're prompted to confirm the addition of this new certificate. Make sure the Keychain pop-up menu is set to "login" and then click Add.**

   At this point, your keychain has been set up correctly. You should see it in Keychain Access, as shown in Figure 5-7.

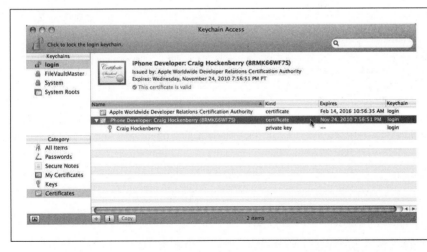

Figure 5-7:
When you're finished with the keychain setup, you'll see something similar to what's displayed here. You'll see your own name after "iPhone Developer" and as the private key, not "Craig Hockenberry".

The key icon in front of your name is particularly important: If it's missing, you won't be able to sign code. If you don't see the icon, you'll need to start over and submit a new certificate request.

It's also important to make sure that the certificate is loaded into the default keychain (shown in bold type). In the figure, it's "login" and selected for viewing. Try reloading the certificate if it's not in the right place.

Provisions

Congratulations, you're halfway toward being able to run your own applications on an iPhone! With your keychain in working order, all that remains is creating a provisioning profile that specifies which devices can run your code.

Identity crisis. If you tried to build your project right now, you'd see an error along the lines of "Code Sign error: a valid provisioning profile matching the application's identifier 'com.iconfactory.SafetyLight' could not be found."

As you learned on page 137, the provisioning profile is a file that associates code signing certificates, application identifiers, and devices. When these three pieces of information are in agreement, you can build and run the application.

So your job now is to start gathering these pieces of information with Xcode and submitting them to the Provisioning Portal.

The new shiny. The first order of business is to determine which devices are going to be used for your development. These devices can be any of the iPhone or iPod touch models.

Tip: You'll find that you want to have multiple devices for your own testing. The performance characteristics of a first-generation iPod touch are very different from those of the current generation. Likewise, older devices don't have the same features as newer ones: Examples are the GPS that wasn't included on the first-generation iPhone and the compass that's available only on the iPhone 3GS. You'll want to test that your application works correctly when these capabilities aren't present.

With a Standard developer account, you're given 100 devices to use for testing. This limit includes devices you use for your internal testing as well as the hardware that your beta testers will be running.

It's important to note that this is a cumulative number: You can't remove one device and replace it with another. Once you enter the unique identifier in the Provisioning Portal, you're stuck with it for the remainder of your yearlong enrollment. That's an important point because the hardware typically gets updated several times each year, and when it does, many of your testers will want to provision their shiny new device. If all of your devices are assigned to the iPhone 3G, you wouldn't be able to develop and test the 3GS model when it's released.

Tip: iPhones are typically updated in the summer around the time of Apple's annual Worldwide Developer Conference (WWDC). New iPods, including the touch models, are usually introduced in the early fall, just before the holiday buying season.

You'll also find that many testers need to change their device identifier throughout the year. When a tester's phone is lost, stolen, or broken, you want to be able to add their replacement to the provisioning profile. The bottom line is that you need to be stingy with devices. Never allocate all of them, and always leave yourself some padding for new and replacement hardware.

Find the Device Identifier. The first step in adding a new device is to determine the Unique Device Identifier (UDID). Each iPhone and iPod touch has a 40-character hexadecimal string that's unlike any other.

Note: Many applications use this UDID for their own purposes. For example, if you need to uniquely identify a request to a web service, you can call the *–uniqueIdentifier* method in the *UIDevice* class and pass the result to the network.

From Xcode, the easiest way to get a UDID is to plug in the device and open the Organizer (Window→Organizer). When asked if you want to use this device for development, answer in the affirmative.

Tip: You'll frequently use the Organizer window, so learn the keyboard shortcut Control-⌘-O.

Once the window is open, select the device in the source list, and the Summary tab appears. If you see a message saying "device could not support development", unplug your phone, restart it, and try again. You can copy the summary's Identifier text to the Clipboard, as shown in Figure 5-8.

Your testers won't be able to get their UDID from Xcode, but they can do the same thing with iTunes once they learn a little trick. After selecting the device, you can click the Serial Number in the summary, and it turns into the identifier. Pressing ⌘-C (on the Mac) or Ctrl-C (on Windows) copies the displayed value to the Clipboard.

Tip: Another good way to collect device identifiers is with Erica Sadun's free *Ad Hoc Helper* app. Your testers can install this simple application from the App Store and run it to automatically send the UDID in an email message.

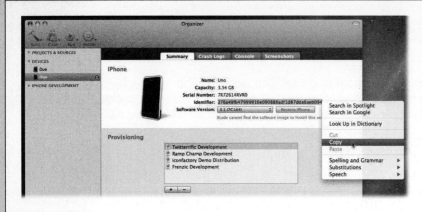

Figure 5-8:
In the Xcode Organizer's source list, you can gather information from your test devices in the Devices section. Here, the Unique Device Identifier is being copied.

Add the device. The next step is to add the device identifiers you've collected:

1. **On the Provisioning Portal web page, click Devices in the left column.**

   The Devices screen shows the current number of devices you have left to register.

2. **Click Add Devices to add your new device.**

   The next screen has two input fields, one for the device name and another for its ID.

3. **Select any name you'd like, and enter the UDID that you want associated to it.**

Tip: You can make your life easier if you follow some simple naming conventions for device IDs, especially when the list gets long.

For example, you can combine the tester's name with the type of device. The name "David Bowman 3G" tells you that Dave uses an iPhone 3G. When he upgrades, you'd add another device named "David Bowman 3GS."

---

4. **After clicking Submit, you see the new device in the list, and the counter is decremented.**

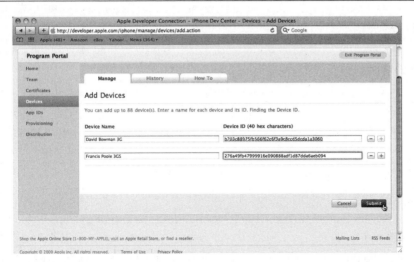

An ID for your apps. The next piece of information you need for the provisioning profile is an App ID. This unique identifier specifies which applications can be run by doing a pattern match. The pattern match can include an asterisk that acts as a wildcard character.

When you were looking at the Info.plist for the flashlight application, you saw that its bundle identifier was defined as *com.iconfactory.FlashlightPro*. This reverse domain name path uniquely defines the flashlight as coming from the Iconfactory. If you've downloaded sample code from the iPhone Dev Center, you'll see that many of the applications have identifiers that begin with *com.yourcompany.*

Say that your domain is *squidger.com,* and your app is named Tiddlywinks Pro. You've defined your bundle identifier in Info.plist as *com.squidger.TiddlywinksPro,* and now have several choices for specifying an App ID:

- ***** matches all applications. You'll be able to build and run any application on your device, including sample code and other third-party apps. But it lacks special powers that the next two options have.

- **com.squidger.*** matches any app that begins with *com.squidger,* including *com.squidger.TiddlywinksPro* and any other future apps your company makes. Use this pattern when you beta test your application. A single provision file that matches all products from your domain makes it easy for testers to try each new creation since they'll already have it installed from the last beta.

- **com.squidger.TiddlywinksPro** is the fully qualified pattern that matches your application exactly. You must use this type of ID if you want to use push notifications or in-app purchases from your app.

For now, take the easiest route and specify your App ID as *. So load up the portal and get on it:

1. **On the Provisioning Portal web page, click App IDs in the left column.**

   The App IDs screen shows the App IDs that have been defined. Over time, the list will grow, and you'll use it when you want to configure push notifications or in-app purchases.

2. **For now, just click New App ID to add your first one.**

   First, fill in the text box under Description to remind you which pattern this is for.

3. **Since you can use this one App ID for building and running all apps, type *All Apps*.**

   Leave the pop-up menu under Bundle Seed ID (App ID Prefix) selected on Generate New (Figure 5-10). Next, enter the pattern used to match the applications under Bundle Identifier (App ID Suffix).

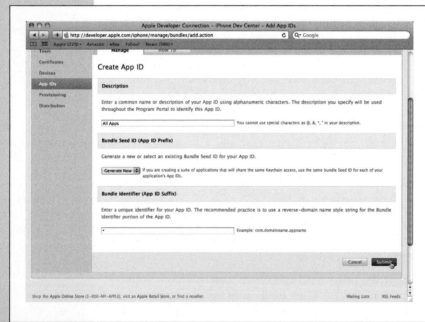

Figure 5-10:
*Here are the settings for your first App ID. A new one is being created using the pattern * that matches all applications.*

4. Type * and click Submit.

After returning to the main App ID page, you see All Apps listed in the description column.

Note: The pattern is prefixed with a randomly generated Bundle Seed ID. This prefix is needed to keep the identifiers unique per developer. They can also be used to share security information between multiple applications by the same developer, but for the most part you can ignore the prefix.

Pull them together with a profile. Good news! You've created all the pieces needed for a provisioning profile—now it's time to assemble them. By now, you won't be surprised to hear that this is going to happen in the Provisioning Portal:

1. **On the Provisioning Portal web page, click Provisioning in the left column.**

   This screen has several tabs. You need to worry only about the Development tab at this point.

2. **Click New Profile to get started.**

   The first thing to do is to give the new provisioning profile a name.

3. **Since this profile is going to be used for general iPhone development, name it** *iPhone Development.* **Turn on all the certificates that are shown.**

   If you're an individual developer, there will be only one certificate. If you're working in a team with other developers, you can add certificates for them, and they'll be listed here, too.

4. **For the App ID, select All Apps. Turn on each of the devices you're going to want to use for your internal development (Figure 5-11).**

   The Select All link is a handy way to choose all devices.

5. **Click Submit.**

   You end up back at the list of provisioning profiles. You'll notice that your new profile's status is In Process. It will take a few moments for Apple's server to generate the file.

6. **Refresh your screen. The status changes to Active, and a Download button appears in the Actions column. Click Download.**

   When complete, you'll have a new *iPhone_Development.mobileprovision* file in your Downloads folder.

At this point, you're done with the Provisioning Portal for a while. Time to switch over to Xcode and add the new file to your build process.

Figure 5-11:
Creating your first provisioning profile, called iPhone Development. This profile pulls together the certificate for Craig Hockenberry, can be used to build and run All Apps, and can be installed on the selected devices.

Xcode setup. You followed a lot of steps in the Provisioning Portal to create the provisioning profile. Luckily, installing it is dead simple. Just drag the *iPhone_Development.mobileprovision* file onto Xcode's icon and you're done!

To verify that it's been installed correctly, open the Organizer in Xcode (choose Window→Organizer). In the Organizer window, open the iPhone Development group in the source list, and find an item named Provisioning Profiles. Click it, and you see all the profiles that can be used by Xcode. You should see *iPhone Development* listed along with an expiration date. If you click the profile, you'll see some additional information about it, as shown in Figure 5-12.

Clean, build, and run. You're now ready to build and install your app on your iPhone. Make sure one of the devices you specified in the provision is plugged in and ready to go. You'll see a green dot next to the device in the Organizer after it's successfully connected to Xcode.

Tip: If the device doesn't show a green dot, select it in the Organizer, and you see a status message in the Summary tab. If you see the message "could not support development", it means that Xcode could not attach to a process on the device. The only way to restart that process is to power-cycle the device.

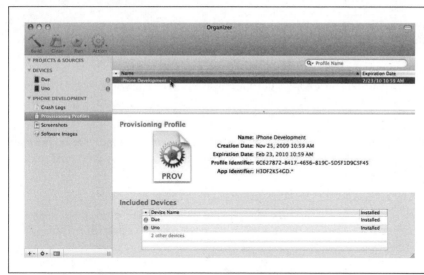

Figure 5-12:
Xcode's Organizer can display the provisioning profiles that are currently installed. The iPhone Development profile was just created using the Provisioning Portal. Note that the expiration date for the profile is about 3 months from the day it was created.

In the main project window, change the Overview menu in the upper-left corner of the window. Up until now, you've been using iPhone Simulator as the Active SDK setting. Now that you have a device that's provisioned and connected, you can select iPhone Device (Base SDK) (Figure 5-13).

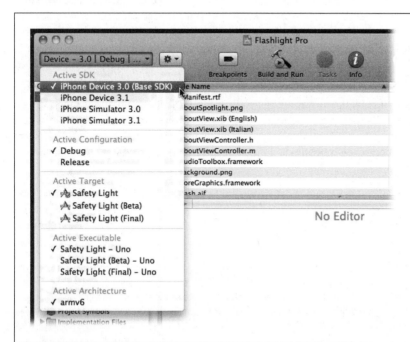

Figure 5-13:
Once you have your iPhone provisioned and connected to your Mac, you can select iPhone Device from the Active SDK menu. The numbers after the name indicate which version of the iPhone SDK you're building against.

Note: Setting the Active SDK for a device causes Xcode to generate code for the iPhone's ARM processor instead of for the i386 processor in your Mac. You can verify this by noting the Active Architecture at the bottom of the Overview menu.

After selecting the Active SDK, clean the project (Build→Clean). This step ensures that you're working on a fresh slate. It removes any code you've generated with Code Sign errors. In the dialog box that appears, make sure both boxes are turned on and click Clean.

And now for the moment every iPhone developer remembers fondly: installing your first app on the iPhone. Choose Build→Build and Run (or press ⌘-Return) to kick things off!

Note: If you see an error about not being able to start the executable, it either means that your iPhone isn't connected or that it needs to be restarted. Also, check that your phone is unlocked so that the home screen is displayed.

You can watch the status line in the lower-left corner as headers are precompiled and source code is compiled. Toward the end of the build process, the code-signing process takes place. When that happens, you see a message box asking if you want "codesign" to access your keychain. Click Allow.

Every time you run an application from Xcode, it checks the device to make sure that a valid provisioning profile is installed. When you dragged the profile onto Xcode's icon, you installed it only in the development environment. The final destination for the provisioning profile is the device, and until it's in place, your application won't run. Luckily Xcode helps you: It prompts you, saying that it can't run your application on the device. When you click "Install and Run", the profile will be installed.

You can verify that the provisioning profile was installed by using the Settings app on your phone. Tap Settings→General→Profiles, and you see a list of installed profiles. You see your iPhone Development profile along with its expiration date. The iPhone OS warns you when a profile is about to expire. You can also use this interface to delete a profile.

Xcode's next step is to copy your application onto the device. You see a status message in the lower-left corner: "Installing Safety Light.app on Kryten…" followed by "GDB: Running…". Now look at your device. Congratulations, you're now an iPhone developer!

Tip: You can connect multiple iPhones and iPods to your Mac as an aid for testing. It's easy—just connect each device to a USB port, and Xcode can use them simultaneously. To control where the app gets installed and debugged, select the appropriate device name for the Active Executable in the Overview menu.

When It Fails

Seasoned iPhone developers will tell you they spend way more time in the Provisioning Portal than they'd like to. The sad truth is that the whole process of signing code is designed to fail over time.

As you've noted along the way, certificates expire, as do the provisioning profiles they're attached to. Even the devices expire because of upgrades or misfortune (consider yourself lucky if you've never discovered a phone in the washing machine!).

Apple has designed this process so that they retain control over the distribution of your application. The result: You're working with a configuration that's definitely *not* "set it and forget it." Worse, you won't remember how you set things up many months ago. The following sections will help you get things back in working order when they break.

Code signing

No one likes a broken build. And if you're seeing errors while signing code, your build is most definitely broken. Time to check your certificates and build settings!

Certificates. The first thing to verify is that certificates haven't expired. The quickest way to do this is to open Keychain Access in your Applications→Utilities folder and to search for *iPhone*. If you see a red *X* on any of your certificates, you'll need to regenerate them. If you saved copies of your previous Certificate Signing Requests, like *iPhone.certSigningRequest* (page 140), you can just resubmit them to the portal. Otherwise, you'll need to generate new requests with Keychain Access.

After your certificates are updated by using the portal, you'll find problems with the provisioning profiles. They have been invalidated because they contain a reference to a certificate that no longer exists. It's not hard to fix: Click the Renew button next to the Invalid status indication.

Once the profile is regenerated, you need to download and install it in Xcode. If other developers on your team are using the same profile, they need to update their development environment as well. Any external testers who are using the profile for beta testing also need to load the new profile on their device.

You can now see how a simple thing like an expired certificate cascades into a lot of work for a lot of people. Good thing you get to be lazy while coding, because you'll need the extra time to manage profiles!

Code signing identities. During the course of your iPhone development, you might see an error message like: "Code Sign error: The identity 'iPhone Developer: Baba O'Riley' doesn't match any valid certificate/private key pair in the default keychain".

This message indicates your project settings include a reference to a certificate that doesn't exist in your default keychain. Check Keychain Access to make sure you see something similar to Figure 5-4.

This problem typically arises when you're working on a project with another developer (either with open source or someone else on your team). If that person changes the code-signing setting to "iPhone Developer: Happy Jack," your certificate won't match and you see the error.

The fix is simple: Make sure that the Automatic Profile Selector is defined in the build settings. Open the target build settings in Xcode (Project→Edit Active Target), and scroll down to the Code Signing section. The Code Signing Identity should be set to Any iPhone OS Device with a value of iPhone Developer. The current match in the keychain shows.

Tip: Remember to check the settings in the target: The target can override any configuration done in the project. Don't fool yourself by looking in the wrong place!

Profile matching. If you've defined a profile using a bundle identifier like *com. mydomain.** and try to build a project that has *com.theirdomain.TheirProduct* in the Info.plist, you'll see an error: "Code Sign error: The identity 'iPhone Developer' doesn't match any identity in any profile."

That's one of the reasons to use the * wildcard for your App ID in the provisioning profile—it's more likely to match one of the certificates in your keychain. Of course in cases where you need to be more specific with your application identifiers, you need to either change the value in Info.plist or generate a new provisioning profile that matches the new identifier.

Device provisioning

You may not be seeing any issues while building your app, but when it won't run on a device, that's a problem! Here are some things to look out for:

Launch abort. When you tap on your application's icon, it starts to launch and you see the image in Default.png for just an instant. Then you're back at the home screen. Bummer. You've just witnessed an expired provisioning profile. The iPhone OS launches the app, determines that there's no valid profile on the device, and then the app quits.

Head to the Settings app, and drill down into General→Profiles. If you see a profile with red text indicating that the profile has expired, delete it. It's no longer of use. Then head to the Provisioning Portal to refresh the provisioning profile so you can reinstall it.

Tip: Deleting expired profiles can help you avoid conflicts between profiles. If you have an expired profile that matches all applications, the OS may find that one first and ignore another valid profile for the application. *Always remove any expired profile* from the device as soon as possible.

It's also possible that the profile wasn't installed properly in the first place. If the Settings app doesn't list the name you've been using for your development, use Xcode's Organizer to install the profile again.

Note: When you send a new beta release to testers, it's likely that you'll hear "the app installed, but won't finish launching" from a few of them. They're having the problems noted above.

Portal confusion. As you add and remove devices in the Provisioning Portal, remember to refresh the profiles in Provisioning. It doesn't happen automatically. Many a developer has updated the device configuration and then immediately downloaded a profile that didn't contain the new information. It's easy to forget that adding a device doesn't magically make it appear in your profiles. You need to click the Edit→Modify link and select each new device in the profile's list.

Similarly, if you delete a device from the portal, any provisioning profiles that use the device are immediately invalidated. Visit the Provisioning section and recreate the affected profiles.

Other problems

Don't be afraid to restart Xcode if you're having problems with code signing or provisioning. For the most part, the development environment handles the management of certificates and profiles without missing a beat. Still, the relationships between the various components are complex, and sometimes Xcode needs a kick in the pants.

Another issue that fools even the best developers isn't related to code signing or provisioning at all. It can be something much simpler—your app was compiled for the wrong version of the iPhone OS.

If you build your app using the 3.1 SDK, your testers can install it on a device with iPhone OS 3.0 installed. When your application won't launch on a device, don't assume that it's a problem with provisioning. Get in the habit of asking your testers which version of the OS they're running before you head off on a provisioning excursion.

Tip: Apps installed via the App Store don't suffer from this issue because the installer checks the version while putting the application on the device. Other installation mechanisms don't perform this check.

You're Now Mobile

As soon as you put your work in your pocket and carry it around with you, you'll begin to look at it differently. Your application is liberated from a desktop development environment and so are your assumptions about its design.

It's extremely important to take your creation into the real world, since your customers will be doing it as soon as they finish downloading it from the App Store. Here are a few things you'll notice immediately:

- **Everything feels slower.** The CPU and network aren't as fast, and that affects the entire user experience.

- **Controls feel smaller.** Those buttons that looked large on your Mac's screen aren't so big anymore.

- **Environmental conditions change.** The UI that looked great in the controlled lighting conditions of your office may not look so good in direct sunlight or harsh fluorescent lighting.

- **Distractions increase.** You'll be using your application while interacting with other people and conditions: Complexity is more difficult to manage.

Over time, you'll find things in your application that break when exposed to the real world. Sometimes it's code, like how your application handles a dropped network connection. Other times it's with design: You need to adjust the contrast and placement of controls.

Either way, you'll find these issues only by leaving the comfort of your work environment.

Ready to Roll

You've got a good handle on the development environment now. Everything is set up the way you want, and you can easily build and test your code. Time to start thinking about the best way to write your application.

It's likely that you've worked with other development environments, so you already know how important it is to come up to speed on the design patterns and best practices that make you an efficient and effective programmer.

That means it's time to dig a little deeper into the Flashlight Pro project!

A Flashlight for Pros

Now that you have your development environment up and running, it's time to start writing code. That's a daunting task when you're dealing with a new language, framework, and platform. That first line of code is hard to write when so many things are so different. Rather than pore through reams and reams of documentation and sample code, you're going to do something a little different now. You're going to look at a complete product that was created by an experienced iPhone developer.

The focus of this chapter is *not* to completely cover every aspect of the programming APIs that are available. Instead, you get a guided tour full of tips and tricks. In essence, you get to sit and watch a pro build an iPhone app. The goal is to teach common patterns and best practices that you can use in your own applications.

A Guided Tour

Luckily, the product to be constructed in this chapter is one that you're already familiar with: a flashlight! The design that you struggled with in Chapter 4 comes to life before your eyes. If you haven't done so already, download the Flashlight Pro project from the Missing CD page (*www.missingmanuals.com/cds*). Everything in this tour references the code in that project.

Tip: Don't be afraid to experiment with this code. In fact, in several areas you can learn a lot by tweaking things. As every good programmer knows, this craft is honed by years and years of trial and error! If you end up breaking the thing irreparably, just download the project from the Missing CD again.

Where to Start?

After your project file is set up correctly, you have a problem every developer faces: blank canvas syndrome. Where do you begin this huge project?

Here are some simple steps to help get you going as quickly as possible:

1. Familiarize yourself with the NIBs and source code created by the template. Some of it you'll want to keep; other parts you'll rename or remove.

2. The application delegate (see the next section) is the first thing that gets executed in your application. It's a natural place for the core of your development to begin.

3. Interface Builder (page 86) is your friend because it lets you see your design almost immediately. The application may not do much because the views created in your NIB file (page 152) can't send actions to the controller, but it inspires you to move forward.

4. Working with Interface Builder also helps you identify which views you need to customize. If you can't get them to look right in the graphical editor, you need to write some code, either by subclassing a standard view (page 67) or by writing your own from scratch.

5. As you start writing view code, you typically encounter the first cases of standard classes that you need to extend with categories or subclasses.

6. Once you have your views in place, start feeding them data. That's a great opportunity to start implementing your models.

7. Throughout the preceding steps, you've been adding instance variables to your controller objects. But at this point, you can really start to pull everything together with the action, update, and notifications between the models and views.

8. As you reach the end of the project, start thinking about localizing the product. Also, clean up the UI with the help of your designer.

You're now ready to begin the tour of the final Flashlight Pro application. You visit the source files in roughly the order they were created. Along the way, this chapter also highlights and discusses important landmarks. So sit back, get comfortable with your copy of the project, and get ready for an excursion through Xcode's Groups & Files list!

Open the Guide Book

Begin your tour by opening the guidebook: Locate the Flashlight_Pro.xcodeproj project file that you downloaded from the Missing CD page, and double-click its icon to open it with Xcode. Make sure that the main Flashlight Pro group and the Application Delegate subgroup are both open, as shown in Figure 6-1.

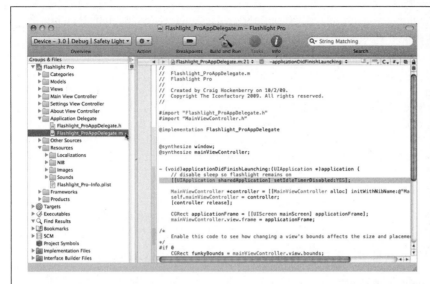

Figure 6-1:
Throughout this chapter, you use the Groups & Files list in Xcode to find the code discussed on these pages. This figure shows Flashlight_ ProAppDelegate.m selected in the list and viewed in the editor. To make the editor fill the right-hand side of the window, use View→Zoom Editor In (⌘-Shift-E). You'll learn more about the first line of code (selected) on below.

As was noted in the list on the previous page, the application delegate is the first place your code is executed, so that's where this tour starts.

Flashlight_ProAppDelegate

There's really only one method that matters in this class's implementation: *-applicationDidFinishLaunching:*. And even though it has only a few lines of code, a lot is going on here.

UIApplication

In Flashlight_ProAppDelegate.m, you see the *UIApplication* class for the first time. One instance of this object is always available, and it allows you to query and configure the state of your application within the iPhone OS.

In this instance, the property that allows the screen to dim and turn off is disabled:

```
[[UIApplication sharedApplication] setIdleTimerDisabled:YES];
```

When you're using a flashlight, you don't want it to turn off until the user wants to turn it off.

The *UIApplication* class also lets you do interesting things like add a badge to your icon on the home screen, change the look of the status bar, show the network activity indicator, and so on.

UIScreen

Another important shared class is *UIScreen*: It contains information about how large the display area is. To determine the size of your view, use:

```
CGRect applicationFrame = [[UIScreen mainScreen] applicationFrame];
```

The reason for writing the code this way, even though you know the exact size of the iPhone's screen (320 × 480, or 320 × 460 with the status bar displayed) is because the size of the screen isn't always what you expect it to be. For example, if you're in the middle of a call or using Internet tethering, the status bar is 40 pixels tall instead of 20 pixels. And if you're in landscape orientation, you're subtracting these values from the width of the screen rather than from the height. It really is best to let Cocoa Touch figure out the current screen size for you.

Tip: Another reason to let the system figure out the dimensions of the screen: A time may come when your app runs on a larger device like the iPad's 1024 × 768-pixel display. You're future-proofing your implementation!

Windows and views

As you continue to look at Flashlight_ProAppDelegate.m, you see where the screen comes to life. It all starts with this code:

```
mainViewController.view.frame = applicationFrame;
```

It should be fairly obvious that the view's size and position (frame) are being set to the entire screen. What's less obvious is that the view object is also getting *instantiated.* That is, because the controller's view is loaded lazily (page 84), the first access to the *view* property of *MainViewController* causes the object to be loaded into memory from the NIB file. As a result, executing this line of code causes the *–viewDidLoad* method to be called in the *MainViewController.*

Having the view appear onscreen is a two-step process. First, you add it as a subview of the window, and then you make the window visible, like this:

```
[window addSubview:mainViewController.view];
[window makeKeyAndVisible];
```

This causes another method in *MainViewController* to be invoked: The code in *–viewDidAppear* runs. Note that this is not the only time this message is sent; it happens anytime the view comes onscreen—when you flip back to the main view after looking at the settings view, for example.

NIB Files: Something to Look At

As you saw in the steps on page 86, it's pretty simple to get a basic interface working using Interface Builder. The UI won't do anything useful, but you can see the controls and other views. It also starts a natural cycle of adding views, creating outlets in the controller, adding actions in the controller, and hooking the actions up to the views.

So what's in the NIB files for Flashlight Pro? Time to start Interface Builder (page 25) and take a look! Just double-click the XIB files listed in Groups & Files.

MainWindow.xib

After you double-click MainWindow.xib, the document and the application window open (Figure 6-2).

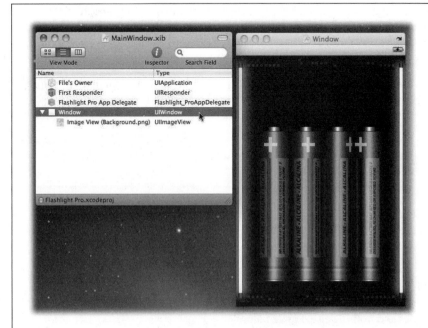

Figure 6-2:
The first NIB file you examine contains a reference to the UIApplication object and an instance of the Flashlight_ProAppDelegate *class. This is also where the window for the application is defined and loaded at launch.*

The first thing you see is that the File's Owner is set to *UIApplication*. As you saw on page 89, this object knows all about your application. It's created automatically when the user launches your app.

A *UIApplication* can have a delegate that is sent messages as the state of the application changes. In this NIB, an instance of that delegate is archived with a type of *Flashlight_ProAppDelegate*. When this file is loaded, you automatically get an instance of this object.

Note that the *UIApplication* object in File's Owner has a delegate that's connected to the instance of *Flashlight_ProAppDelegate*. That ensures that your delegate will know when the application state changes. Without this important connection, the –*applicationDidFinishLaunching:* message would never be sent, and you'd end up with a window without a view. When you configured the Main NIB File in the target settings as MainWindow, you told Cocoa Touch to automatically load this file at launch. That setting is an important part of the application bootstrap process!

Tip: If you ever have a problem with this first NIB file appearing, make sure that the *NSMainNibFile* key in Info.plist has the correct value.

When you open the disclosure triangle next to the Window instance, you see that it has a subview that's a *UIImageView*. That's why you see the pictures of the batteries when you double-click the Window to view it in Interface Builder.

Subviews stack atop one another. This image view with the batteries sits at the bottom of the stack. You might be wondering why you don't see these batteries when you launch your application. Because you stacked something on top of it with the *–applicationDidFinishLaunching:* implementation:

```
[window addSubview:mainViewController.view]
```

The view at the very bottom of the stack is still there, though. And that's why you see it every time you flip between the main view and the settings view.

Note: This stacking of views is called the *view hierarchy*. You're not limited to just one stack, either. Each view can have its own stack of views, allowing for an arbitrarily deep and wide collection of user interface elements.

One final connection of interest in this NIB file: *Flashlight_ProAppDelegate* has an outlet set for the window. When this object is loaded from the file, the instance variable is automatically connected so any message sent to *window* will be delivered to the *UIWindow*.

If this connection is broken, the window instance variable is *nil,* and the messages to *–addSubview:* and *makeKeyAndVisible* are discarded. You end up with a picture of batteries and nothing else. Power failure!

MainView.xib

When you open this NIB file and double-click Light View, what you see is similar to what you see when you run the application (Figure 6-3). That's good, because this is the main controller view.

Figure 6-3:
Here are the contents of the NIB file for the application's main view. The view that displays the light is named Light View, and it has subviews for the toolbar and status information. You can also see that MainViewController is responsible for managing the contents of Light View.

File's Owner

Note that the File's Owner is set to *MainViewController*. Generally, the owner of a NIB file is the object that caused it to be loaded. As you saw on page 157, *UIApplication* was the File's Owner, since it was responsible for loading MainWindow.xib automatically.

When you execute this code in –*applicationDidFinishLaunching:*, *MainView-Controller* loads MainView.xib:

```
MainViewController *controller = [[MainViewController alloc]
    initWithNibName:@"MainView" bundle:nil];
```

That makes *MainViewController* the File's Owner for MainView.xib.

Note: When you create a new *UIViewController* subclass by using the New File template, you have an option to create it "With XIB for user interface". When you turn on this option, the File's Owner is configured automatically.

If you need to change the class for some reason, you can do it using the Identity Inspector in Interface Builder. Just select the File's Owner in the list of objects, press ⌘-4, and select a new class from the drop-down menu.

When there's a mismatch between the class used for File's Owner and the one used in your code, a crash with *NSInternalInconsistencyException* can occur in –*_loadViewFromNibNamed:bundle:*. If the class setting looks right in the NIB, also check that the view outlet is set correctly.

Having the controller as the main object in the NIB file also gives you access to the outlets and actions that are defined in the class' header file. To see these connections, select the *MainViewController* in the list of objects. Then press ⌘-2 to bring up the Connection Inspector, and hover your mouse over each outlet and action. The connected view is highlighted in the Light View window.

Views upon views

When you open the disclosure triangle to the left of Light View, you see an Info View and a Toolbar. If you open the Toolbar, you see "Flasher", "Disco", "SOS", and "Settings" separated by "(Flexible Space)".

The view for the light has two subviews: one for the information and another for the toolbar. The toolbar, in turn, has seven subviews consisting of four buttons and three separators (all from the *UIBarButtonItem* class).

Sometimes it's easier to navigate via this list than it is to select items from the Light View window. If you click Disco, for example, you can press ⌘-1 and quickly see that the button's image is ToolbarDiscoStart.png, or press ⌘-2 to see that it sends the *toggleDisco* action. You can also use the arrow keys to move around in the list of objects to quickly check all your object's properties in the current inspector.

Custom views

This NIB file contains two custom views: *LightView* and *IFInfoView*. When you're creating custom views, you're often faced with a "chicken and egg" problem. You want to place an instance of the view in the NIB *before* you have an implementation of the class.

To work around this situation, you can create an instance of a *UIView* in your view hierarchy. You can then use this generic view for placement and other basic configuration. Then, after you've implemented class, you can use Identity Inspector (⌘-4) to change the Class Identity from *UIView* to the name of your new view.

You'll see the implementation of the *LightView* and *IFInfoView* classes later in this section (page 169).

View Sizing

One of Interface Builder's inspectors has been noticeably absent so far: the one that lets you set the size of a view (Figure 6-4). You can access this panel by choosing Tools→Size Inspector or by pressing ⌘-3.

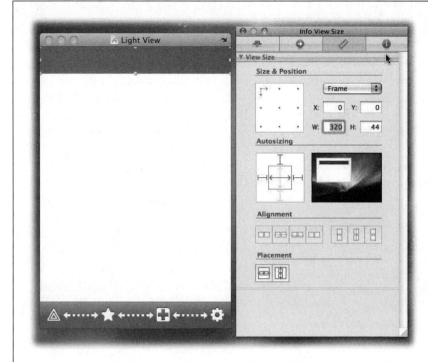

Figure 6-4:
You can see the Info View and its settings in the Size Inspector panel. The Size & Position values define the origin of the view along with its width and height. The Autosizing control shows four anchors and two stretching arrows. The Info View is anchored to the left, top, and right edges and will stretch horizontally when the parent view's size changes. The toolbar has similar settings, except it's anchored to the bottom of the view rather than to the top.

Sometimes you want to specify a view's frame with parameters. Info View is a good example: You want to make sure that it's in the upper-left corner of the parent view and that its height is set to 44 pixels (so it matches the height of the toolbar).

By specifying X and Y coordinates of 0,0, you ensure that the view is placed at the upper-left origin. Likewise, a width of 320 and height of 44 set the initial width and height.

Autoresizing

Setting the dimensions is only a start because a view's size can change as your application runs. Whether it's by doubling the height of the status bar or by changing a device orientation, the application's screen area will vary. Luckily, Cocoa Touch provides an elegant mechanism for automatically resizing views. Called *autoresizing*, it's available for any view and is propagated through the hierarchy.

Experiments in Resizing

The "autosizing" preview gives you a hint of the resizing effect, but a better way to check your resizing configuration is to change the view's orientation. Even if your app doesn't support a landscape mode, clicking the arrow in the top-right corner of a view window causes automatic resizing to be applied.

Here are some simple steps in Interface Builder that should help you understand this complex feature:

1. Open the Light View in MainView.xib. In the window that appears, click the orientation change icon in the upper-right corner.

2. The view will turn on its side, and both the Info View and Toolbar are repositioned correctly, as shown in Figure 6-5. That's autoresizing in action! Click the orientation arrow again, and the view returns to a portrait orientation. The views return to their original positions thanks to autoresizing.

3. Now open the Light View Attributes Inspector (⌘-1) and uncheck Autoresize Subviews. After you click the orientation control, note how the toolbar falls off the bottom of the screen because the view still thinks the height of the light view is 460 pixels. Likewise, the information view's width remains 320 pixels even though the screen is now 480 pixels wide.

4. Since you don't want to save these changes to your NIB file, use Interface Builder's File→Revert menu

item. After clicking the Revert button, the last saved version of the file is reloaded.

5. To see how the view stretching works, select Info View from the list of objects. Bring up the Size Inspector (⌘-3), and click the width arrows in the middle of the autosizing control. (No red arrows should be in the middle of the black box.)

6. Click the orientation control, and you see that the width of the information view remains at 320 pixels.

7. Leave the orientation in landscape. Now you're going to adjust the anchor point for the Toolbar, so select it from the object list and open the Size Inspector.

8. Change the anchor at the bottom of the view to the top. Click the red anchor at the bottom of the black box to turn it off. Then click the anchor above the box to turn it on.

9. Now, when you change the orientation back to portrait, the toolbar moves to the middle of the window. That's because the toolbar is maintaining the same distance between the top edges of the toolbar and light view.

Feel free to play around with the anchor and stretching controls for all of the views to get a feel for how things work. Just make sure not to save changes to the MainView. xib file, or you're going to have a flashlight that behaves very strangely when it resizes!

When you select Light View and inspect its attributes (⌘-1), you see that Autoresize Subviews, in the Drawing section, is turned on. When the size of this view changes for whatever reason, both the Info View and Toolbar are resized automatically.

Figure 6-5:
When you click the orientation change icon in the upper-right corner of the Light View window, the view switches between portrait and landscape. Use this icon to quickly test the autoresizing behavior of your views. In this figure, you'll see that both the Info View and Toolbar have a new width, but remain anchored to the same edges of the screen.

So what does it mean to resize a subview? A good question that leads to another control in the Size Inspector! Select Toolbar from the list, and press ⌘-3 to bring up the inspector. You'll see an "autosizing" section in the palette. The control here tells the system how to compute new view dimensions when resizing occurs. The outer edge shows four anchor points. The inner square shows how the view will be stretched vertically and horizontally. When a view can be stretched, the anchor points define the gap between the edge of the view and the superview. If the view can't be stretched, then the anchors define a relative position to maintain. Each of these six controls affects the final view size; if you hover over the control, a preview is displayed.

The toolbar is configured to stretch its width and is anchored on all sides except the top. Since no pixels are between the edge of the toolbar and the light view, any automatic resizing keeps the control attached to the full width of the view. Because the toolbar is not anchored at the top, the position of the view shifts up if the size of the screen changes.

SettingsView.xib

When you open SettingsView.xib and double-click Settings View in the object list, you see something very similar to what is displayed in the application (Figure 6-6). With the exception of the customized graphics in the slider controllers and the items in the picker view, you see the same views.

Try Resizing on for Size

After you build your application, you should always test its resizing in the Simulator. You can change the orientation using the Hardware→Rotate Left and Rotate Right menus. To get a double-height toolbar, use Hardware→Toggle In-Call Status Bar.

A common problem is seeing your toolbar pushed down onscreen. When this occurs, check that autoresizing is turned on for the main view and that the anchor at the bottom of the toolbar is set.

When you add a new *UIView* as a subview to the main view, it's anchored on all four sides and stretches in both width and height. This maintains relative placement within the super-view: If you have a view that needs to be positioned against the edge of the screen, you need to update the defaults.

Figure 6-6:
With the exception of the custom graphics for the slider controls and the content of the color picker, the SettingsView.xib document looks very much like what you see in the application. The Settings View is composed of many subviews.

Click the disclosure triangle to the left of the Settings View in the object list, and you see that this view is composed of many subviews.

Opacity

The order of these views is important. The first subview listed, Light View, is drawn at the bottom of the stack. Then the two image views, the picker view and others, are drawn on top of each other in order.

By default, views are opaque. That's the fastest way for the iPhone to draw, since it doesn't have to use an alpha channel to composite (blend) one view on top of another. Only the topmost view is drawn. Likewise, if you're sure that the view completely fills its rectangle, you can configure it to not clear the area before drawing. These configuration parameters are listed as Opaque and Clear Content Before Drawing in the Attributes Inspector.

In views that need to draw quickly, turning off transparency and not clearing before drawing can speed things up considerably. This is typically an issue when you're drawing table views and want scrolling to be as smooth as possible. This simple settings view has a limited number of views, and the drawing is done infrequently, so transparency doesn't affect drawing performance significantly.

Opacity is a very important part of making sophisticated iPhone user interfaces. Views that are transparent make the job much simpler because you're able to layer effects to produce a final result. The first example is each slider view. The icon images used for the minimum and maximum values have a transparent background. The knob and slider images are also transparent. These elements combine to create a slider that can be drawn on any background.

Another example is the combination of the opaque Light View with a semitransparent Background Image View to produce the preview. The image view uses the image SettingsBackground.png to draw the view. When you open that image with the Finder, you see that it has a fully transparent hole and a partially transparent penumbra. This image is drawn on top of the opaque view of the light. When the Light View's color changes, it forces the view layers above to recompose and produce a new image onscreen. This produces the effect of light being cast on a surface without any complicated drawing.

Tip: You can open any file in an Xcode project in the Finder by Control-clicking (or right-clicking) the file name in the Groups & Files list and choosing Open With Finder from the shortcut menu. You can also choose the "Reveal in Finder" item to open the folder that contains the file.

For an image file, Open With Finder will display the graphic in the Preview application. If you're viewing a lot of files with transparency, you may want to adjust the Window background color using Preview→Preferences→General.

The sliders

One final thing to note in this NIB file is the slider configuration. The Brightness Slider is set to update continuously, while the Speed Slider is not. This configuration is done using the Continuous checkbox in the Attributes Inspector (⌘-1).

Sometimes you want to update your controller (and models) as the user drags the slider. Other times that continuous update can cause performance or visual problems. When the Speed Slider is updating continuously, the flashing becomes erratic, so it was turned off.

AboutView.xib

Hey—this NIB isn't a part of the original plan! You're a developer—you know that plans change. As you were developing the product, it became apparent that you needed a place to promote your business. As you were beta testing, you also found that it was really handy to have a place to display the beta version number. And once you saw how easy it was to add a controller and views, it took only a short time to implement this new About View (Figure 6-7).

Figure 6-7:
This figure shows an About View that contains credits, promotional information, and the application's version number.

Tip: This new NIB file might also be a place for your humble author to add few more cool examples, too.

Most of the view consists of labels with text describing the product. One of those views, the Version Label, is connected to the *–versionLabel* outlet in the controller. This way, the version number can be updated automatically. You'll see how that works when the tour passes by the *AboutViewController* class (page 204).

Another interesting view in this NIB is the Close Button. Take a look at its size in the inspector: It's a whopping 320 × 460 pixels, and it covers the entire screen!

That button lets the user tap anywhere on the screen and send the *–close* action to the controller. It shows that buttons don't necessarily have to show any visible text or image to be effective. You can use transparent buttons to act as hit detectors on an image, for example.

Finally, there's the Spotlight Image View. It's a transparent image that looks kinda boring now, but wait until you get to the controller (page 205). That view is going to move and groove!

Refine the Look

Now that you've seen the NIB files, it's time to take a look at how some of the custom views were created. There's a custom view for the light source and another for information displayed at the top of the screen.

LightView

This view is modeled after a physical light. Objects that emulate the real world are usually easier to design, since the correct behavior is already something you're familiar with. They also tend to make your code easier to read.

You see this tendency a lot in Cocoa Touch. For example, the *UISwitch* control manages its state by using a property named *on*. Which of these snippets of code is more readable?

```
mySwitch.on = YES;
```

or

```
mySwitch.state = YES;
```

The properties of the view let you change the state of the light (on, off, or a single pulse). You can also set the brightness and color of the light. To model the physical property of the light, you can also set an envelope where attack and release variables control the amount of time the light takes to turn on and off.

Delegation

Delegation is not just for system classes. The light uses this design pattern to notify another object that taps have occurred in the view. In this application, that object is *MainViewController*. When the tour reaches the controller code (page 194), you see how it acts as the delegate.

When you look at LightView.h, you see a forward declaration of the protocol:

```
@protocol LightViewDelegate;
```

This declaration is needed because there's a chicken-and-egg problem between the delegate and the view. Both the class and protocol definition need each other.

Using this forward declaration, the delegate instance variable is defined as:

```
id <LightViewDelegate> delegate;
```

This variable tells the compiler that the delegate can be any object as long as it conforms to the *LightViewDelegate* protocol. (Remember that *id* is a generic object of any class.)

The full protocol declaration follows the *LightView @interface*:

```
@protocol LightViewDelegate <NSObject>

@optional
- (void)lightView:(LightView *)view singleTapAtPoint:(CGPoint)tapPoint;
- (void)lightView:(LightView *)view doubleTapAtPoint:(CGPoint)tapPoint;

@end
```

The two messages sent by the *LightView* are *–lightView:singleTapAtPoint:* and *–lightView:doubleTapAtPoint:*. Both are tagged as *@optional,* so it's up to the delegate whether it implements these methods. If *@optional* is not specified, the delegate is required to provide the methods.

Enumeration

C enumerations are used throughout Cocoa Touch as a way to express multiple options in code. The light view uses an enumeration to control the state of the light. One example in UIKit is the *UIViewAnimationCurve* enumeration. The enumeration's naming and values follow a simple pattern—a concatenation of the class name, the property name, and a unique option name. So the animation curve types are *UIViewAnimationCurveEaseInOut, UIViewAnimationCurveLinear,* and so on.

Yes, it's wordy, but very easy to read in code:

```
[UIView setAnimationCurve:UIViewAnimationCurveLinear];
```

Note: This standard naming pattern is very useful with automatic completion in Xcode. Once you know the rules, finding these long names is easy.

For example, you can type *UIViewAnim* and press Esc to display a list of suggestions. You know to type the initial text because you're working with a *UIView* and the *animationCurve* property.

The light view follows this pattern with its own enumeration:

```
typedef enum {
    LightViewStateOff = 0,
    LightViewStateOn,
    LightViewStatePulse
} LightViewState;
```

which is used in the instance variable definition:

```
LightViewState state;
```

and in the property:

```
@property (nonatomic, assign) LightViewState state;
```

The controller can change the state of the light by using the enumeration:

```
lightView.state = LightViewStatePulse;
```

Class extensions for private methods

In Chapter 2, you learned about Objective-C categories as a way to extend an existing class (page 38). A variation called *class extensions* lets you extend the interface of your own class. Even though you're the one writing the class and you can define anything you want in the interface in the implementation, there's an advantage to extending your class's interface instead: It lets you hide your internal methods.

For example, the *@interface* in LightView.h shows methods that any other class can use to communicate with the object. The problem is that some methods are only used internally, and exposing them in the interface is both confusing and dangerous.

The solution is to define another *@interface* in your implementation. This *@interface* looks very much like a category, only it's anonymous because there is no name in the parenthesis:

```
@interface LightView ()

- (void)updateBackgroundColor;
- (void)endPulse;
- (void)handleSingleTap;
- (void)handleDoubleTap;

@end
```

The methods in this *@interface,* like *–updateBackgroundColor,* are extending the original one and must therefore be implemented in the implementation below. As a result, you can call *[self updateBackgroundColor]* in LightView.m as often as you'd like, but if another object tries to send a message like *[lightView updateBackground-Color],* the compiler will complain.

Note: Other object-oriented languages use the term *private methods.* Many Objective-C programmers use this term, too.

Catching view property changes

As you've seen previously, the *@property* and *@synthesize* declarations can save you lots of time writing your accessor code. There's a problem, though: You don't know when this accessor code is being executed.

Most of the time, that's not a problem. In a view, however, a property update often means that some presentation of the data must adapt. The view's *color* property is an example. When another object alters the color, you need to update the view to show the change.

To update the view, you need to write out the accessor and add your own code:

```
- (void)setColor:(UIColor *)newColor {
    if (color != newColor) {
        [color release];
        color = [newColor retain];

        [self updateBackgroundColor];
    }
}
```

You can't be lazy all the time, and view property changes are one of those times.

Tip: If you're doing your own custom drawing for a view, the only extra code you typically need is:

```
[self setNeedsDisplay];
```

The *–setNeedsDisplay* method marks the view as being dirty and causes redrawing to occur automatically using the *–drawRect:* method. You'll see this technique in action with *IFInfoView* (page 174).

Note also that the drawing won't be done immediately. Instead, multiple drawing requests will be coalesced into a single update in the next event-processing cycle. If you try to circumvent this by calling *–drawRect:* directly, it's less efficient and unpredictable. Let the system decide when it's time to draw.

View animation

Cocoa Touch lets you animate some view properties, and that's how it makes things look so cool. But more importantly, it's the way you give users subtle clues for the actions they perform.

Here are a few view properties and the kinds of effects you're used to seeing in iPhone applications:

- **alpha.** Fading a view in or out when a user edits or otherwise changes the state of the application.

- **frame.** Moving smoothly between two points to add or remove information onscreen.

- **transform.** Rotating, scaling, and flipping data by altering an affine transformation matrix.

- **backgroundColor.** Gently changing the background color of a view to indicate a state change.

Animation is done in *blocks.* A block is a group of animation property changes that act in unison. Changes to view properties between the beginning and end of the block appear simultaneously and provide a smooth experience for the user. Animation blocks also simplify your code.

Tip: Nothing stops you from changing the alpha, frame, and background color of the view simultaneously. Just remember that the best animations are the ones that act as subtle clues. Graduates of the Sledgehammer School of Design only impress themselves, not customers.

The light view animates just one property: the background color that's used as the light bulb in your virtual flashlight. All of the work is done in the private *–update-BackgroundColor* method.

You begin an animation block with the *+beginAnimations:context:* class method for *UIView*:

```
- (void)updateBackgroundColor {
    [UIView beginAnimations:@"updateLightView" context:NULL];
```

The *updateLightView* is used by delegate methods to uniquely identify an animation. With sophisticated transitions, more than one animation can occur at a given time, so being able to pick the right one is important.

Each animation lets you set various options. For the light view, if an animation is already in progress, the current color will be the starting point for the new color:

```
    [UIView setAnimationBeginsFromCurrentState:YES];
```

Other properties for view animations are a repeat count, whether it reverses when completed, and several others that control the timing.

The light view uses one of the most common properties: the animation duration. If the light is on, the amount of time specified in the *envelopeAttack* instance variable is used to control how long it takes to change the *backgroundColor* from its current color to the new one:

```
    if (self.state == LightViewStateOn || self.state == LightViewStatePulse)
    {
        [UIView setAnimationDuration:self.envelopeAttack];
        [self setBackgroundColor:self.color withBrightness:self.brightness];
    }
```

If the light's state is off, then the *envelopeRelease* variable is used as the *animation-Duration* while setting the background color to black.

```
    else {
        [UIView setAnimationDuration:self.envelopeRelease];
        self.backgroundColor = [UIColor blackColor];
    }
```

Once you've configured the view animation, you use the *–commitAnimations* class method to start things in motion.

```
    [UIView commitAnimations];
}
```

The result is that the light view smoothly transitions between colors and state. The view looks and acts like a real light bulb, thanks to view animation.

You'll see more of these animation blocks when the tour reaches the category and the *SettingsViewController* (page 201). Once you're comfortable with how they work, they can add quite a bit of polish to your application. Just remember you can have too much of a good thing.

IFInfoView

The other custom view in the Flashlight Pro project is the one used to display status at the top of the screen. Unlike the *LightView,* which relied on a *backgroundColor* property to display itself, *IFInfoView* does custom drawing. In this section, you also learn how to set up the view and name it properly.

Name

Why is *IF* at the beginning of the view class name? Couldn't you just call it *InfoView,* like *LightView?*

Yes, you could. But you're a lazy programmer. And you have a vision of the future….

Someday, this information view could be useful in other projects. Lots of applications need to show status from time to time, and this code is a great way to do that. Chances are that you'll want to reuse this code.

And if you try to put this class in a project that already has an *InfoView,* you're going to have a name conflict. Since Objective-C doesn't have namespaces, you've got to fake it.

Just as frameworks use prefixes like *NS* and *UI* (page 33) to avoid collisions, you can add your own prefix to make your classes unique. This code comes from the Iconfactory, so it uses the standard *IF* prefix. There's an implicit guarantee that no other classes from this company will use that name more than once.

With this system, the Iconfactory can use the code in its own projects, and you can also use it in yours. Sure, namespaces would be a more elegant solution, but in practice, this gentleman's agreement works rather well.

Tip: If you ever find yourself with a name conflict for classes, make sure to look at Xcode's Refactor tool. It makes it easy to change the name of a class throughout the entire project (including in NIB files). You can also use it to change the names of instance variables and to change other aspects of your classes. To use this feature in Xcode, select the class's name in the source code editor and choose Edit→Refactor.

View setup

You can create any object in two ways: either by using a NIB file or by using an *–init* method. That's especially true with view classes.

In both cases, you likely have some code to set instance variables to a valid state. In the case of this view, an information icon is displayed by default. You could perform this setup in the *–initWithFrame:* and *–awakeFromNib* methods, but as your views become more complex and you add instance variables, this duplicated code becomes difficult to manage and keep in sync.

The solution is to create a *–setupView* method that can be called in both instances:

```
- (void)setupView {
    type = IFInfoViewTypeInformation;

    self.backgroundColor = [UIColor clearColor];
}

- (id)initWithFrame:(CGRect)frame {
    if (self = [super initWithFrame:frame]) {
        [self setupView];
    }
    return self;
}

- (void)awakeFromNib {
    [self setupView];
}
```

This ensures that the *backgroundColor* is completely transparent and that the default type is set no matter whether the view is created via code or from being read from NIB.

Drawing in code

The heart and soul of this class is in the *–drawRect:* method. Whenever the *IFInfoView* needs to be updated and appear onscreen, this method is called.

In most cases, this drawing is initiated using the *–setNeedsDisplay* method in the *UIView* parent class. For example, when the *type* property is changed, you want to see a new icon, so you mark the view as needing display:

```
- (void)setType:(IFInfoViewType)newType {
    if (newType != type) {
        type = newType;

        [self setNeedsDisplay];
    }
}
```

Now that you know what brings about drawing, it's time to look at the code in the *–drawRect:* method.

Tip: Before you start writing a bunch of custom drawing code with –*drawRect:*, try using a basic view hierarchy. This class could have been implemented as *UIImageView* with a background image and two subviews. The subviews would draw the text using a *UILabel* and the icon using another *UIImageView*.

Color filling

One of the most basic operations is a standard color fill. First you pick the color you want to fill with:

```
UIColor *fillColor = nil;
if (self.type == IFInfoViewTypeAlert) {
    fillColor = [UIColor colorWithRed:0.6f green:0.0f blue:0.0f alpha:1.0f];
}
else {
    fillColor = [UIColor colorWithRed:0.0f green:0.0f blue:0.0f alpha:0.60f];
}
```

If the information type is an alert, a solid 60% red is chosen; otherwise a black with 60% opacity is used. Once you've selected the color, you must call its –*set* method so that it will be used in subsequent drawing operations:

```
[fillColor set];
```

The only thing left to do is to fill the rectangle being drawn:

```
UIRectFill(rect);
```

Gradients

It's time to go from one of the simplest drawing operations to one of the most complex—you're going to draw a gradient on top of that solid color.

To do so, you use *CoreGraphics*. This framework contains the entire low-level API used for drawing on the iPhone. You already encountered this framework when you used any definition that begins with *CG*, notably *CGRect* (page 158).

To begin drawing the gradient, you need an array of colors:

```
UIColor *startColor = [UIColor colorWithRed:1.0f green:1.0f blue:1.0f
    alpha:0.45f];
UIColor *endColor = [UIColor colorWithRed:1.0f green:1.0f blue:1.0f
    alpha:0.10f];
CGColorRef colors[] = { [startColor CGColor], [endColor CGColor] };
CFArrayRef colorsArrayRef = CFArrayCreate(NULL, (const void **)colors, 2,
    NULL);
```

The color that's used for the gradient is white; only the opacity varies. It starts at 45% and ends at 10%. (That's 55% and 90% transparency.) Both values go into an array of *CoreGraphics* color references.

Note: You're using an array because you can specify as many colors as you'd like in a gradient. You can also specify the points at which the colors are applied.

The next step is to use these colors along with a color space reference, and to create a reference to the gradient specification:

```
CGColorSpaceRef colorSpaceRef = CGColorSpaceCreateDeviceRGB();
CGGradientRef gradientRef = CGGradientCreateWithColors(colorSpaceRef,
    colorsArrayRef, NULL);
```

Now that you have the gradient defined, you can draw it in the current graphics context. The graphics context defines the entire 2-D drawing environment: UIKit provides the *UIGraphicsGetCurrentContext()* function so you're able to share that environment with the system code that's drawing views:

```
CGContextRef contextRef = UIGraphicsGetCurrentContext();
CGContextDrawLinearGradient(contextRef, gradientRef, CGPointMake(0.0f, 0.0f),
    CGPointMake(0.0f, CGRectGetMidY(rect)), 0);
```

The gradient is being drawn from the top of the view to the middle of its rectangle.

Tip: Many convenience functions allow you to create and query the Core Graphics data structures. In the above example, you see *CGPointMake()* and *CGRectGetMidY()*. There are also functions like *CGRectMake()* and *CGRectGetMinX()*.

The last step is very important. The references created by Core Graphics allocate memory, so you need to explicitly release that memory (page 43).

```
CFRelease(colorsArrayRef);
CGGradientRelease(gradientRef);
CGColorSpaceRelease(colorSpaceRef);
```

Since view drawing tends to be done repeatedly, memory leaks while doing these screen updates can seriously impact your application.

Whew! That gradient was tough. It's going to get a bit easier next.

Blended drawing and lines. Another common type of drawing is *blending*. It's often used to make a part of an image lighter or darker to indicate some kind of state (like selection). Another use, like the one used in the info view (page 175), is to highlight the edges and provide additional contrast.

The top of the rectangle is lightened with a 20% white using this code:

```
lineRect = CGRectMake(0.0f, 0.0f, rect.size.width, 1.0f);
[[UIColor colorWithRed:1.0f green:1.0f blue:1.0f alpha:0.20f] set];
UIRectFillUsingBlendMode(lineRect, kCGBlendModeLighten);
```

This works like the *UIRectFill()* function you saw on page 176, but it adds a blending mode as a second parameter. The same drawing function is used at the bottom of the rectangle. Note that when the blend mode is *kCGBlendModeDarken,* you set a black color with 30% alpha. This code also shows a simple trick to draw horizontal and vertical lines. You just define a rectangle whose height or width is 1 pixel.

A path to greatness

So what do you do if the line you want to draw isn't vertical or horizontal? Or what if you want to draw something other than a rectangular area? Core Graphics to the rescue!

Drawing a "line" with this framework is a fairly complex operation. And for good reason: The mechanism that is used is not limited to just lines. You can draw circles and other ellipsoids. Or you can draw complex shapes derived from Bezier paths.

This drawing operation is based on a *path.* A path can be stroked to generate the outline of a shape or filled to generate a solid. Paths are generated incrementally: You move to a point, add a line to another point, append a curve, and so on.

Here's some code that draws a line at the bottom of the rectangle:

```
lineRect.origin.y = rect.size.height - 1.0f;
[[UIColor colorWithRed:0.0f green:0.0f blue:0.0f alpha:0.30f] set];
UIRectFillUsingBlendMode(lineRect, kCGBlendModeDarken);
```

To do the same thing using a path, you'd first create it:

```
CGMutablePathRef pathRef = CGPathCreateMutable();
```

Then add the two end-points of a line:

```
CGPathMoveToPoint(pathRef, NULL, 0.0f, rect.size.height - 0.5f);
CGPathAddLineToPoint(pathRef, NULL, rect.size.width, rect.size.height - 0.5f);
```

You can do any kind of drawing at this point. If you're drawing a rectangle with rounded edges, you might use *CGPathAddArc().* To draw a Bezier curve, you use *CGPathAddQuadCurveToPoint().* You can also add rectangles and other geometric shapes to the path.

Once you finish constructing the path, you close it:

```
CGPathCloseSubpath(pathRef);
```

The next step is to get the graphics context set up for drawing. Set the blend mode along with the color and width of the line:

```
CGContextSetBlendMode(contextRef, kCGBlendModeDarken);
[[UIColor colorWithRed:0.0f green:0.0f blue:0.0f alpha:0.30f] set];
CGContextSetLineWidth(contextRef, 1.0f);
```

Then add the path to the current graphics context. The context can then use the path for subsequent drawing operations:

```
CGContextAddPath(contextRef, pathRef);
```

And finally you draw the line by telling the graphics context to stroke the path:

```
CGContextStrokePath(contextRef);
```

Make sure to clean up afterwards, too. The memory used by the path must be released, and the blending mode should be returned to its default state:

```
CGPathRelease(pathRef);
CGContextSetBlendMode(contextRef, kCGBlendModeNormal);
```

This is a lot more work than the three lines of code you started with, but it provides a lot more flexibility. When you need to draw something more than a line or rectangle, this is the way to go!

Half a pixel matters

Make sure to pay attention to the coordinates used when drawing. Many a developer has been tripped up by the precision in Core Graphics. When you draw a 1-pixel line, half of the line is above the specified coordinate, and the other half is below (Figure 6-8).

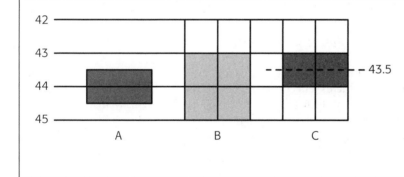

Figure 6-8:
A line drawn at integral coordinates won't appear as you expect. When you specify 44.0 as a vertical coordinate, half the line will be drawn above and half below (A). The line is spread across 2 pixels, losing color definition and becoming blurry (B). The solution is to subtract half the line width and to draw at 43.5 (C).

For example, if you specify a line with a width of 1.0 and a horizontal coordinate of 44.0, the top of the line that's drawn is at 43.5, and the bottom is at 44.5. Since Core Graphics smoothes the line using anti-aliasing, it splits the single line of color across 2 pixels. This results in a blurry line that loses color definition.

To compensate, the heights used while defining points have half the line width subtracted, resulting in a line that spans the 44th pixel exactly.

Note: This blurry drawing can also occur when you compute view rectangles. As with lines, rectangles that are off by half a pixel will look wrong, and it won't be clear why it's happening.

For example, if you use the *CGRectGetMidY()* function to compute the midpoint of a rectangle that has an even height, you'll get a fractional result. In cases like this, use the *floorf()* math function to clamp the result as an integral value. In almost all cases, rectangle origins and sizes should use integers even though they are defined as floating-point values.

Text drawing

The next thing to draw in the information view is the text of the message. To match the style of the toolbar, the string will be drawn with a 1-pixel shadow to give the appearance of it being etched into the curved surface created by the gradient.

Before you can draw the text, you need to figure out its size. To do this, you need the font it's drawn with:

```
UIFont *font = [UIFont systemFontOfSize:17.0f];
```

This code gets you Helvetica at 17 points. *UIFont* also has methods to get bold or italic versions of the system font. You can also retrieve a font by name if you want to draw custom text.

Once you have the font, you need to specify the area used to draw the text. Since you want the message to fit into the width of the rectangle after a 44-pixel margin for the icon, you define the size constraints like this:

```
CGSize constrainedSize = CGSizeMake(rect.size.width - iconOffset,
    9999.0f);
```

The second parameter is a large value because you're interested in the computed height, so there's effectively no constraint in that dimension. To compute the height of the text, use a UIKit category for *NSString* that includes the *–sizeWithFont:constrainedToSize:* method:

```
CGSize textSize = [self.message sizeWithFont:font
    constrainedToSize:constrainedSize];
```

This, in turn, is used to find a new rectangle that centers the text in the height of the view:

```
CGRect textRect = CGRectMake(iconOffset,
    (CGRectGetMidY(rect) - (textSize.height / 2.0f)) - 1.0f,
    rect.size.width - iconOffset,
    textSize.height);
```

Now that the math is done, you can actually draw the text. If you're paying close attention, you might wonder why that *–1.0* appeared in the computation. The reason is that you're going to use a trick to draw the shadow for the text: You draw it twice. The first time it's drawn, you use a dark color and offset 1 pixel above the normal text.

Tip: If you want to add a shadow to complex shapes or other kinds of drawing, you can use the *CGContextSetShadow()* function with the current graphics context.

Round Your Rects

Paths can also be used to clip while drawing. A common use of this is to display round corners when drawing images that have square corners.

With this technique, you construct a path using lines connected by arcs at the corners:

```
#include <math.h> // for M_PI

#define MIN(A,B)    ((A) < (B) ? (A) : (B))


CGFloat DegreesToRadians(CGFloat degrees)
{
    return degrees * M_PI / 180;
}


CGPathRef RoundedRectPath(CGRect rect,
  CGFloat radius) {
    CGMutablePathRef pathRef =
      CGPathCreateMutable();

    radius = MIN(radius, 0.5f *
      MIN(CGRectGetWidth(rect),
      CGRectGetHeight(rect)));
    CGRect insetRect = CGRectInset(rect,
      radius, radius);

    // top-left corner
    CGPathAddArc(pathRef, NULL,
      CGRectGetMinX(insetRect),
      CGRectGetMinY(insetRect), radius,
      DegreesToRadians(180.0),
      DegreesToRadians(270.0), false);

    // top-right corner
    CGPathAddArc(pathRef, NULL,
      CGRectGetMaxX(insetRect),
```

```
      CGRectGetMinY(insetRect), radius,
      DegreesToRadians(270.0),
      DegreesToRadians(360.0), false);


    // bottom-right corner
    CGPathAddArc(pathRef, NULL,
      CGRectGetMaxX(insetRect),
      CGRectGetMaxY(insetRect), radius,
      DegreesToRadians(0.0),
      DegreesToRadians(90.0), false);


    // bottom-left corner
    CGPathAddArc(pathRef, NULL,
      CGRectGetMinX(insetRect),
      CGRectGetMaxY(insetRect), radius,
      DegreesToRadians(90.0),
      DegreesToRadians(180.0), false);


    CGPathCloseSubpath(pathRef);

    return pathRef;
}
```

After adding this path to the graphics context, you can call *CGContextClip()*, and it will be used to mask while drawing the image:

```
CGPathRef pathRef =
  RoundedRectPath(rect, 8.0f);
CGContextAddPath(context, pathRef);
CGContextClip(context);
[myImage drawInRect:rect];
CGPathRelease(pathRef);
```

To create dark black with 50% transparency:

```
[[UIColor colorWithWhite:0.0f alpha:0.5f] set];
```

And here's the code that draws the shadow characters in the view:

```
[self.message drawInRect:textRect withFont:font];
```

The second pass at drawing the text adjusts the origin by moving it down 1 pixel, setting the color to white, and using the same method to draw the characters:

```
textRect.origin.y += 1.0f;
[[UIColor whiteColor] set];
[self.message drawInRect:textRect withFont:font];
```

Pretty simple and clever, huh?

Note that the –*sizeWithFont:constrainedToSize:* method is also used when the message for the view is changed. If the text is longer than one line, the computed size is used to adjust the height of the view.

Image drawing

There's one last thing to draw in the view: the icon image. The first step is to read the image from the application resources:

```
UIImage *image = nil;
if (self.type == IFInfoViewTypeAlert) {
    image = [UIImage imageNamed:@"InfoViewAlert.png"];
}
else {
    image = [UIImage imageNamed:@"InfoViewInfo.png"];
}
```

The *UIImage* class method +*imageNamed:* locates the image file within your application bundle and caches it in memory.

The next step is to compute the placement of the image within the draw rectangle. The –*size* method for *UIImage* gives you the image dimensions, which are then used to find a rectangle that's vertically centered in the view:

```
CGRect imageRect = CGRectZero;
imageRect.size = [image size];
CGRect drawRect = CGRectMake(CGRectGetMidX(imageRect),
    CGRectGetMidY(rect) - CGRectGetMidY(imageRect),
    imageRect.size.width, imageRect.size.height);
```

To get the image to appear in the view, all you need to do is send the –*drawInRect:* message to the *UIImage* instance:

```
[image drawInRect:drawRect];
```

You're done drawing the view and learned a lot in the process. Congratulations!

Tip: When you're drawing a view, it's always a good idea to base all geometry on the rectangle you are passed. As you've seen, view sizes can change with or without your knowledge. If you make assumptions about sizes or hard-code values, you're setting yourself up for failure.

In some cases a constant value makes sense: The *iconOffset* acts as a margin between the image and informational message.

Making Cocoa Touch Your Own

As you saw in Chapter 2, categories in Objective-C are very powerful mechanisms for extending existing classes (page 38). Flashlight Pro has three categories. Two are used to create and set colors with a brightness value, and a third is used to conceal a view onscreen.

UIColor+Brightness

The first category you're going to look at is *UIColor+Brightness*. This category takes some code you found on the Internet, wraps it in Objective-C, and integrates it with the existing *UIColor* class.

Naming conventions

The first thing you notice about the category is the name of the source file. Since these categories tend to be shared among projects, a naming convention is important to avoid conflicts. You'll find that your own categories get a lot of reuse, and you can also find a lot of great open source ones.

In the convention used here, the names are derived from the class being extended and the name of the category. For this category, the *@interface* is defined as:

```
@interface UIColor (Brightness)
```

That makes the name *UIColor+Brightness*. If the category name had been "(Chockolicious)," the files would be named *UIColor+Chockolicious* instead.

Note: Some developers use a dash as a separator instead of a plus sign. The separator isn't really important; it's the unique combination of the original class and the category that matters.

Wrapping it up

If you look at the *@implementation,* you see a lot of C code that converts between the RGB and HSV color spaces. (The source code in this section comes from an ACM paper by Alvy Ray Smith and Eric Ray Lyons. To see it for yourself, search the Web for *converting RGB to HSV written in C.*)

This code illustrates a couple of important points. First, Objective-C and standard C get along really well. The code in the algorithm is unchanged. The only modifications were stylistic, mainly to make the names and formatting fit in the new environment.

The other significant thing is how this fairly complex code is made more accessible via the Objective-C wrapper. Without looking at any documentation, you can guess what this code does:

```
CGFloat brightness = [[UIColor orangeColor] brightness];
[mySlider setValue:brightness];

UIColor *backgroundColor = [[UIColor redColor] colorWithBrightness:0.6f];
[myView setBackgroundColor:backgroundColor];
```

When you decide to use a category, one line of code is most important:

```
#import "UIColor+Brightness.h"
```

This lets the compiler know about the additional methods and link them into the final binary. If you forget to do this, you see "may not respond to" warnings when you build your app.

UIView+Brightness

This simple *UIView* category uses the *UIColor* category you just looked at to add a brightness parameter when setting the background color for the view. Nothing prevents one category from using another: They really do behave like the original class as long as you remember to import the *@interface*.

UIView+Concealed

As a part of your design, you wanted the information view and the toolbar to go offscreen at times. Information doesn't need to be shown indefinitely, and the toolbar needs to be hidden when the flashlight is locked.

You could have implemented these features by adding methods to *IFInfoView* and creating a subclass of *UIToolbar*. But you're smarter than that!

A new property

You could have used the existing *hidden* property for *UIView* to toggle the visibility of the views. But that's not very elegant and doesn't give the user a clue that the information is short lived. Adding a bit of animation to the process of hiding the view solves both problems.

This *UIView* category adds a *conceal* property to any view. The beauty of this approach is when you find a bug or want to add a new concealing feature. You only need to modify the code once, not in two or more views.

The *UIView+Concealed* category adds a new property to a view. You animate the view on- or offscreen using the *–setConcealed:* method. Likewise, you can check if the view is visible with the *–isConcealed* message.

Note: Categories can't add instance variables to a class. In most cases, properties refer to one of these variables, but it's not a requirement. Properties only define a contract for accessing some data internal to the class.

This category uses the existing frame property (and *UIView* instance variable) to determine if the view is onscreen or off.

For example, a controller could toggle the view's concealment using this new property:

```
myView.concealed = ! myView.isConcealed;
```

All the controller needs is this single line of code at the top of the implementation:

```
#import "UIView+Concealed.h"
```

View coordinates

When you conceal a view, you're changing its coordinates within the super-view. Before you can get to the code that moves the view around, you need to spend a little time understanding the view coordinate system used in Cocoa Touch.

You were actually exposed to this coordinate system at the beginning of this tour. While learning about *Flashlight_ProAppDelegate,* you saw this line of code:

```
CGRect applicationFrame = [[UIScreen mainScreen] applicationFrame];
```

If you were curious, you might have also looked at the documentation for *UIScreen* and seen only one other method in that class: *–bounds.* You've also seen the term *frame* repeatedly in this chapter. Simply put, the *bounds* define the coordinate system where subview frames are placed (Figure 6-9).

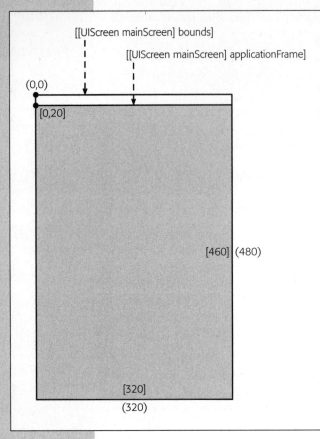

[[UIScreen mainScreen] bounds]

[[UIScreen mainScreen] applicationFrame]

(0,0)

[0,20]

[460] (480)

[320]

(320)

Figure 6-9:
*The bounds of the iPhone's screen, shown with
values in parenthesis, have a zero origin with a
width of 320 and a height of 480. The frame is
placed within these bounds and has an origin
and height adjusted by 20 pixels. Frame values
are shown in square brackets.*

Here's an example using the *UIScreen –applicationFrame* and *–bounds*. As you know, the screen for the iPhone is 320 × 480 pixels. So *UIScreen* returns a rectangle with these sizes when you query the *–bounds* method.

Your application, however, can't use all these pixels because of the status bar. The *–applicationFrame* method returns a frame rectangle that's based on the bounds. That frame is used to place your first view in the window.

It's unlikely that you're just going to have one view in your iPhone application. And that's fine, because when you set a view's frame, you're also setting its initial bounds. When you set your first view's frame size to 320 × 460, you also set the bounds.

If you're thinking that this is duplicating data, it is. But for a good reason: The new bounds have a zero origin that is used for placing subviews. These subviews don't need to know anything about the frame of the superview. The situation is a little confusing, so Figure 6-10 provides a picture that's hopefully worth a thousand words.

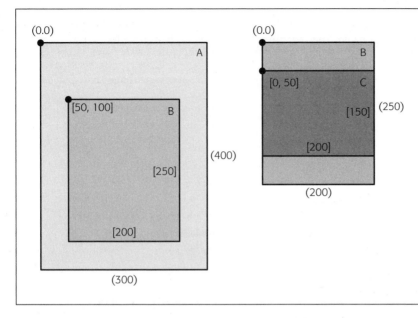

Figure 6-10:
This picture shows how three views, A, B, and C, are placed onscreen using a frame and bounds coordinates. View A has bounds of 300 × 400, and subview B has a frame size of 200 × 250 with an origin at 50, 100. When you specify the placement of subview C, you use view B's bounds of 200 × 250.

Note that view C in the diagram has no idea what the frame of the superview B is set to. Nor does it know what the bounds of view B are. Having two sets of coordinates, frames, and bounds lets view C worry only about where its frame is placed within superview B's bounds.

Note: For the most part, there's a one-to-one relationship between bounds and frame coordinates. One pixel in the bounds is the same as one pixel in the frame. Why would you want to mess with that?

Changing the bounds can scale a view: If you halve the size of the bounds and don't change the size of the subview frame, the effective size of that subview is also halved. Similarly, changing the origin of the bounds moves all subviews. To see the effect, change the *#if* from 0 to 1 in *Flashlight_ProApp-Delegate*:

```
...
mainViewController.view.frame = applicationFrame;

CGRect funkyBounds = mainViewController.view.bounds;
funkyBounds.origin = CGPointMake(-10.0f, 10.f);
funkyBounds.size.width /= 2.0f;
funkyBounds.size.height /= 2.0f;
mainViewController.view.bounds = funkyBounds;
...
```

This test code also makes it easier to see how view concealment works.

More view animation

In the *LightView* class, you saw some basic animation to change the background color. In this category, you see more animation code. This time it's used to move the view's frame so that it slides smoothly on- and offscreen.

Now that you're an expert with view frames and superview (page 162), you can see how the *–isConcealed* property is implemented:

```
CGRect superviewBounds = self.superview.bounds;
CGRect viewFrame = self.frame;

if (viewFrame.origin.y >= superviewBounds.size.height) {
    // view is below the superview's frame
    result = YES;
}
else if (viewFrame.origin.y < superviewBounds.origin.y) {
    // view is above the superview's frame
    result = YES;
}
```

If the view's frame is above or below its parent's bounds, then it's considered concealed. Similarly, *–setConcealed:animated:duration:* uses the *superviewBounds* to compute a new *viewFrame* that is either on- or offscreen depending on the concealed parameter.

But what makes the frame move around smoothly? It's this little bit of code at the end:

```
if (animated)
{
    [UIView beginAnimations:@"updateConcealedView" context:NULL];
    [UIView setAnimationDuration:duration];
}
self.frame = viewFrame;
if (animated)
{
    [UIView commitAnimations];
}
```

Many methods within UIKit use an animated parameter. Animation is cool, but when you're initializing views, it's not the best solution. Having startup code that takes 1/3 of a second will make the app feel awkward, so calling this code in your view controller's *–viewDidLoad* method solves the problem:

```
[self.myView setConcealed:YES animated:NO];
```

Progressive methods

The method definitions in this class category also show another common technique used by Objective-C developers. Methods are defined progressively. The method that does all the work is *–setConcealed:animated:duration:*, and you can use this method if you want to specify each parameter explicitly.

In most cases, however, you're fine with the default settings, so you can just send the *–setConcealed* message (or use the *concealed* property). And, as you saw above, you can use the *–setConcealed:animated:* method for quick initialization.

Tip: This approach is similar in concept to default parameters in C++.

A new feature

Do you remember how using a category for concealment (page 185) let you easily add new features to all your views? The *–concealAfterDelay:* method is a perfect example.

After adding the methods to manage the view's placement, you realize you're doing a lot of work in your view controllers to manage the concealment. Calling *–setConcealed* all the time is a pain.

Two methods and eight lines of code later, you have this solution:

```
- (void)conceal {
    self.concealed = YES;
}

- (void)concealAfterDelay:(NSTimeInterval)delay
{
    [NSObject cancelPreviousPerformRequestsWithTarget:self
        selector:@selector(conceal) object:nil];
    [self performSelector:@selector(conceal) withObject:nil afterDelay:delay];
}
```

When you send a view the *–concealAfterDelay:* message along with a delay, the view is automatically concealed after that period has elapsed. Your view controller can now do this:

```
self.myView.concealed = NO;
[self.myView concealAfterDelay:3.0f];
```

This controller relies on a very cool method in *NSObject* that lets you invoke a selector and pass it an object. These two lines of code are functionally identical:

```
[myObject myMethod:myParameter];
[myObject performSelector:@selector(myMethod:) withObject:myParameter];
```

Which, in and of itself, is pretty neat. You can use this technique to dispatch messages at runtime, not just at compile time. But it gets even more impressive because you configure how long to wait before sending the message by using the *afterDelay* parameter:

```
[self performSelector:@selector(conceal) withObject:nil afterDelay:delay];
```

This sends the *–conceal* message after 3 seconds or whatever you've specified with the delay variable. When that happens, the concealed property is set to NO, and the view begins to conceal itself.

Since this method may be called multiple times before the time elapses, this line of code cancels any pending messages:

```
[NSObject cancelPreviousPerformRequestsWithTarget:self
    selector:@selector(conceal) object:nil];
```

Generally, if you're going to perform a selector at a later time, you should cancel any existing ones first. If there aren't any, the cancel message is essentially ignored.

Fashion Your Models

Now that you've ventured through the views of Flashlight Pro, it's time to look at the models that provide data to the application.

You might be thinking, "Data? For a flashlight?"

Yes! Data comes in many forms: The models for the flashlight let you keep track of the light's settings and build a state machine that controls the SOS flasher. It's not data in the traditional sense, but it's still information that you separate from your controllers.

SOSModel

When you're sending an SOS signal, you're dealing with a stream of dots and dashes that change over time. In essence, you're working with a state machine. Since one of a model's functions is to manage internal logic and behavior, it makes the distress signal a good candidate for this kind of object.

The model's interface is simple: A *unitInterval* property lets you define how long a dot or a dash takes (dashes are defined as being three times as long as a dot). There is also a method that lets you query the current element.

You start the state machine running by invoking the *–send* method.

Notifications

After you start the state machine running, how do you find out when to change from a dot to a dash? You could poll the element parameter, but you know that polling for value updates is terrible to do in an event-driven environment. And Cocoa Touch is very much an environment driven by events.

The solution is to rely on notifications. As you saw on page 97, notifications let you distribute information to objects in a structured manner. So that's what you do with the SOS model. Here's the notification name as it's defined in the *@interface*:

```
extern NSString *const SOSModelElementChangedNotification;
```

which has a corresponding definition in the *@implementation*:

```
NSString *const SOSModelElementChangedNotification =
    @"SOSModelElementChangedNotification";
```

As the state machine advances, the notification is posted so that any interested view controllers can update their views:

```
- (void)update {
    [self advance];

    [[NSNotificationCenter defaultCenter]
        postNotificationName:SOSModelElementChangedNotification object:self];

    [self performSelector:@selector(update) withObject:self
        afterDelay:[self duration]];
}
```

When this tour passes by the *MainViewController* class, you'll see how it handles the notification.

Note: This class design is very similar to how you'd handle a network refresh for data loaded via the Internet. Instead of *–send,* you'd implement *–refresh,* which loads data asynchronously using an *NSURLConnection*. When the data finishes loading over that connection, you'd send a *refresh did finish* notification so any registered view controllers could update their views using the new data.

Private property

Class extensions can also be used to define private properties. You get the advantage of having Objective-C create your accessor methods, and you don't have to expose internal data.

The *SOSModel* uses an integer instance variable to maintain state. Surprisingly, this variable is named *state* and is defined in the *@interface*:

```
    NSUInteger state;
```

Notice that no *@property* definition is in the header file. You'll find it in a class extension just above the *@implementation*:

```
@interface SOSModel ()

@property (nonatomic, assign) NSUInteger state;

...

@end
```

At this point you can use the *state* property just as if it had been defined externally. And you can be assured that a client of your class won't mess around with your internal data.

FlashlightModel

You can use the flashlight model to persist the application's state. The current brightness, flashing speed, color, and light pattern are available via properties and methods. Whenever you change one of these values in the model, it's immediately saved in the user's settings database.

The model also provides a small database of colors. For example, a *UIColor* instance can be retrieved along with a localized name for the user interface.

Read-only properties

As you saw on page 191, in some cases, keeping other objects from snooping around in your data is a good idea. Also, in some cases, you don't mind exposing your state, but you don't want it to be modified. In these cases, you'd use a read-only property. The *currentColorIndex* in the *@interface* does just that:

```
@property (nonatomic, readonly) NSUInteger currentColorIndex;
```

Now you have a bit of a problem. You can read the property, but you can't write to it. If you try to use *self.currentColorIndex = 0,* you get a compiler error.

Class extensions to the rescue! You can override the property definition in the implementation file:

```
@interface FlashlightModel ()

@property (nonatomic, assign) NSUInteger currentColorIndex;

...


@end
```

This lets your *@implementation* write to the property, but anyone using your *@interface* will only be able to read it.

Class statics

On page 100, you saw how to use class static variables to implement a singleton pattern. The flashlight model contains a slight variation on this theme. The class static is used to hold an array of color names and values:

```
+ (NSArray *)colors {
    static NSArray *colors = nil;
    if (colors == nil) {
        colors = [[NSArray alloc] initWithObjects: ...
            ...
```

The first time the +*colors* class method is used, the array is allocated and retained for the lifetime of the application.

Tip: You can use class statics to speed up parts of your application. If you have an object that's accessed frequently, and it takes a while to load the object, you want to minimize the time you spend instantiating it.

A static variable with a *UIFont* or *UIImage* can be initialized once and used repeatedly while drawing your view. These cached objects can help improve your scrolling speed in a *UITableView.* Just remember that you're never getting back the memory used by the cache, so use this technique only for small objects.

Class statics are also used for the keys that store and retrieve the color information:

```
static NSString *nameKey = @"name";
static NSString *colorKey = @"color";
```

This approach allows the compiler to check the key type and catch typos. If you use @"*name*" to write the data in the hash table, and a misspelled @"*nane*" to read it, you're going to be doing some head scratching. By using *nameKey,* you avoid this potential problem.

User defaults

Cocoa Touch provides a standard mechanism for storing a user's settings. It's called *NSUserDefaults* and it acts as the backbone for this model.

For example, when you send the –*setSpeed:* message to the model, the current speed value is updated and then stored as a floating-point value in the standard user defaults by using the –*setFloat:forKey:* method:

```
- (void)setSpeed:(float)newValue {
    if (newValue != speed) {
        speed = newValue;
        [[NSUserDefaults standardUserDefaults] setFloat:speed
            forKey:speedKey];
    }
}
```

When the model is initialized, the user defaults are checked using –*objectForKey:* to see if a speed value has been set. If that method returns a *nil* value, a default setting is assigned. Otherwise, the current setting is read with –*floatForKey:.*

```
- (id)init {
    if ((self = [super init])) {
        NSUserDefaults *userDefaults = [NSUserDefaults standardUserDefaults];
...
        if (! [userDefaults objectForKey:speedKey]) {
            speed = 0.75f;
            [userDefaults setFloat:speed forKey:speedKey];
        }
        else {
```

```
                    speed = [userDefaults floatForKey:speedKey];
            }
    ...
        }
        return self;
    }
```

If you had just called *–floatForKey* without checking for its object first, you would have gotten back a value of zero and not known if that was because it was the user's choice or a nonexistent setting.

Since settings are stored in property lists, the object classes that can be stored are *NSArray, NSDictionary, NSString, NSData, NSDate,* and *NSNumber* (integer, floating-point, and Boolean values).

Strings in user interfaces

One of the reasons for keeping a small database of colors is to associate a color value with a human-readable name. Remember that not all humans read the same language you do. That's why a simple string like *@"White"* is wrapped with a localization macro:

```
    NSLocalizedString(@"White", nil)
```

Whenever you're using a string value that's going to show up in a user interface, you should localize it using this technique. How these strings get assigned and processed will be described when the tour passes by the Localizable.strings file.

Pull Yourself Together

Now that you've seen the views and models in the application, it's time to examine how controllers make everything work together. They're the glue that pulls your flashlight together.

You've already had a preview of the controllers: There's one controller for each of the NIB files that contains a view (page 86, for example). Of course, the basic controller created by Xcode doesn't have any instance variables or actions, nor does it have any of the logic specific to the application.

So hop back in the bus and let the tour continue....

MainViewController

The design of the main flashlight view is simple by design. This leads to a controller that uses both models, but only has three view outlets. Two of the views are custom *UIView* subclasses. As with most controllers, the interesting part is how it interacts with the views and models.

Crack Open a Simulator

The Simulator is a powerful debugging tool. But it has a problem: Every time you build and run a new version of your app, it's copied into a new folder. This folder uses a long string of hexadecimal numbers (in the GUID format) and is hard to locate.

And you want to locate it while debugging, because that's where your application's documents and settings are stored. When working on models, it's often helpful to locate the backing store for the data. Whether the file contains a SQLite database, a property list, or even a plain-text file, it's important that you can verify what's been written.

Here's a short shell script that you can use to help locate your application data:

```sh
#!/bin/sh

if [ -z "$1" ]; then
    echo "usage: $0 <app> [ Preferences |
      <document> ]"
else
    base=~/Library/Application\ Support/
      iPhone\ Simulator/
    apps=Applications
    app=`ls -1td "$base/"*"/$apps/"*"/$1.
      app" | head -1`
    if [ -n "$app" ]; then
      dir=`dirname "$app"`
      if [ "$2" = "Preferences" ]; then
```

```sh
        open "$dir/Library/Preferences"
    else
        open "$dir/Documents/$2"
      fi
    fi
  fi
```

Put this script somewhere on your PATH, name it *opensim*, and make it executable. Once it's installed, you can use it to find the most recent application loaded in the Simulator and open the Preferences, Documents, or a named document in the Finder.

For example, to open the current Documents folder for the Safety Light application, just do this:

```
$ opensim "Safety Light"
```

No documents are in that folder, but if there were, you could open them with:

```
$ opensim "Safety Light" fleshy.db
$ opensim "Safety Light" awesome.plist
```

You can easily get to your preferences folder with:

```
$ opensim "Safety Light" Preferences
```

If you double-click the com.iconfactory.SafetyLight.plist file that's displayed in the folder, the flashlight's current settings are displayed.

Note: A copy of the *opensim* script is located in the top level of the Flashlight Pro project, so you don't have to type it in.

Notifications

Any object can register to receive notifications, but they're used most frequently by controllers. Controllers typically have the logic necessary to respond to the changed conditions for the device, models, or other parts of the application.

The main view controller registers itself for notifications from *UIApplication* and the *SOSModel*:

```
- (NSObject *)initWithNibName:(NSString *)nibNameOrNil bundle:(NSBundle *)
    nibBundleOrNil {
...

        [[NSNotificationCenter defaultCenter] addObserver:self
            selector:@selector(elementChanged:) name:SOSModelElementChanged
            Notification object:nil];

        [[NSNotificationCenter defaultCenter] addObserver:self
            selector:@selector(resignActive:) name:
            UIApplicationWillResignActiveNotification object:nil];
        [[NSNotificationCenter defaultCenter] addObserver:self
            selector:@selector(becomeActive:) name:
            UIApplicationDidBecomeActiveNotification object:nil];

...

}
```

And just as importantly, it removes itself as an observer when the view controller object is freed:

```
[[NSNotificationCenter defaultCenter] removeObserver:self
    name:SOSModelElementChangedNotification object:nil]
[[NSNotificationCenter defaultCenter] removeObserver:self
    name:UIApplicationWillResignActiveNotification object:nil];
[[NSNotificationCenter defaultCenter] removeObserver:self
    name:UIApplicationDidBecomeActiveNotification object:nil];
```

The *–resignActive:* and *–becomeActive* methods in the controller are invoked when you use the lock switch on the device. These methods start or stop the flasher or disco lights. It's a simple, but good, example of how a change in device state results in a change of application state.

The notification defined by the *SOSModel* sends the *–elementChanged* message every time the state of the model changes. This method, in turn, turns the light view on or off—an example of how new data is propagated to a view.

Sounds

The *–elementChanged* method also plays a dot or dash beep sound:

```
if (sosDotBeep != 0 && sosDashBeep != 0) {
    if (element == SOSModelElementDot) {
        AudioServicesPlaySystemSound(sosDotBeep);
    }
    else {
        AudioServicesPlaySystemSound(sosDashBeep);
    }
}
```

Tip: If you want your sound to play along with a vibration, use the *AudioServicesPlayAlertSound()* function instead.

Here's where these *sosDotBeep* and *sosDashBeep* sounds come from: When the flasher was started with the *–startSOS* method, these sounds were loaded into memory with this code:

```
OSStatus result;
NSString *dotPath = [[NSBundle mainBundle] pathForResource:@"Dot"
    ofType:@"aif"];
result = AudioServicesCreateSystemSoundID((CFURLRef)[NSURL
    fileURLWithPath:dotPath], &sosDotBeep);
if (result != kAudioServicesNoError)
{
    sosDotBeep = 0;
}
```

First the sound file is located within your application's bundle. The *bundle* is a group of code and resources for your application. All of the files in Xcode's Resources group are moved into the main bundle.

To extract one of these files from the bundle, you specify the file using its resource name (*@"Dot"*) and its type (*@"aif"*). The resulting *dotPath* variable is then converted into a file URL and passed to the *AudioServicesCreateSystemSoundID()* API. If the call succeeds, the *sosDotBeep* variable can be used to play the sound.

Note: Some resources, notably images, have their own mechanism for being read from the bundle. To extract an image, you'd use *[UIImage imageNamed:@"xyz.png"]*.

Since the sounds use a fair amount of memory, they should be loaded only when needed. If you need a sound for only a short period, you may want to use *AudioServicesDisposeSystemSoundID()* to free its storage.

The standard Xcode templates aren't able to access these sounds APIs; you need to add a new framework before you can compile the code above. If you open the Frameworks group in Xcode, you see that the *AudioToolbox.framework* has been added.

Being a delegate

The *LightView* class defined a protocol for a delegate. The light view uses the delegate to pass on information about taps within the view. This controller uses the delegate methods to change the brightness and lock the view. Its first step to becoming a delegate is to adopt the protocol in the interface:

```
@interface MainViewController : UIViewController <LightViewDelegate,
    UIAlertViewDelegate, UIActionSheetDelegate>
```

The *LightView* in MainView.xib has an outlet that's connected to the *MainView-Controller*, so the view knows who will be acting as the delegate. All that's left to do is to implement the *–lightView:singleTapAtPoint:* and *–lightView:doubleTapAtPoint:* methods.

The delegate for the single tap checks to see if the SOS signal is running using one of the controller's state variables. It also checks whether the toolbar is concealed because the display is locked. If either of these conditions is met, the tap is ignored. After determining the current brightness level, a new brightness is chosen, and the state change is displayed in the info view using localized strings.

A double tap also checks if the SOS signal is running and will ignore the tap if it is. Otherwise, the state of the toolbar's concealment is toggled, and the info view is updated.

One view, two identities

Careful observation will show that the *LightView* object is connected to two outlets in MainView.xib. There's only one instance of *LightView,* but it's used for both the *infoView* and view properties. Wouldn't one instance variable be enough?

Yes and no.

If you use just one instance variable, you have to use the view that's required by the *UIViewController.* And if you use just that variable, you find that it being defined as a *UIView* base class creates a lot of extra effort. If you try to use this code, you get a compiler error:

```
self.view.state = LightViewStateOn;
```

So you'd be forced to write this code to make the compiler happy:

```
((LightView *)self.view).state = LightViewStateOn;
```

Besides going against your lazy tendencies, that code is just plain ugly! If you define a second outlet for the object with the correct type, you can do this:

```
self.lightView.state = LightViewStateOn;
```

The only real cost for this code is to make sure that you release the instance in your *–viewDidUnload* and *–dealloc* methods. No additional memory is used, and your code is much more readable.

Alert views

When you were looking at the protocols adopted by the main view controller, you might have noticed that it also adopted the *UIAlertViewDelegate* protocol. Meet your first alert view. The first time you start the flasher, a warning is displayed that flashing lights can cause seizures. The goal of this app is to make people's lives a bit safer, not put them in danger!

Alert views are common in iPhone apps for situations like these. If an error occurs or an important choice must be made, it's a great interaction mechanism. Don't use them as status indicators, though—folks get annoyed with too many onscreen alerts popping up.

The alert is displayed with the following code. First the model is checked to see if the user has accepted the risks of running the flasher:

```
if (! self.flashlightModel.flashingAccepted) {
```

If they haven't accepted, an alert view object is created with a title, a message, and labels for the buttons being displayed in the view:

```
UIAlertView *alertView = [[UIAlertView alloc]
    initWithTitle:NSLocalizedString(@"FlasherWarning", nil)
        message:NSLocalizedString(@"FlasherWarningMessage", nil)
        delegate:self
        cancelButtonTitle:NSLocalizedString(@"Cancel", nil)
        otherButtonTitles:NSLocalizedString(@"FlasherWarningButton
            Start", nil), NSLocalizedString(@"FlasherWarningButton
            Learn", nil), nil];
```

Since your controller is a *UIAlertViewDelegate,* you set it as the delegate. Note also that the strings are localized. If you're going to warn someone about the dangers in your app, you need to do it in the language they understand best.

Sending the alert view a *–show* message causes the alert to appear onscreen. You then release the object because it's no longer needed and end the action method:

```
[alertView show];
[alertView release];
return;
}
```

But then what? You're asking for user input, and nothing is being returned by the alert view. Is some kind of telepathy going on here?

No, that feature won't be available until Apple releases the iPhone 9GSXY. In the meantime, you need to remember that you're a *UIAlertViewDelegate.* And the following code will be called when the user taps a button:

```
- (void)alertView:(UIAlertView *)alertView
    clickedButtonAtIndex:(NSInteger)buttonIndex {
    if (buttonIndex == 0) {
        // Cancel
    }
    else if (buttonIndex == 1) {
        // Start Flasher

        self.flashlightModel.flashingAccepted = YES;
```

```
        [self startFlasher];
        self.flasherRunning = YES;
    }
    else if (buttonIndex == 2) {
        // Learn More
        [[UIApplication sharedApplication] openURL:[NSURL
            URLWithString:@"http://en.wikipedia.org/wiki/Photosensitive_
            epilepsy"]];
    }
}
```

Each button in the alert has an index. The Cancel button is the first, followed by a button to start the flasher, and another to learn more about photosensitive epilepsy.

Note: The "learn more" button shows how you can open a web page in Safari. First you convert a string to a URL, and then you pass it to *UIApplication*. Your application will be closed, and Safari will open with the specified page:

```
NSURL *URL = [NSURL URLWithString:@"http://chocklock.com"];
[[UIApplication sharedApplication] openURL:URL];
```

Action sheets

Your original plan was to have a "disco mode." In fact, a lot of the variable names reflect this feature. But one day your marketing expert realizes that disco dancing and safety don't really have much in common. Other kinds of lights might be more appropriate.

The result is that this feature morphs into one that lets users select from a list of lights. There are emergency and alert lights for dangerous situations, and the disco light is the option for folks who want to do a little dance, make a little love, and get down tonight. To make the choice, you give your app an action sheet!

Action sheets are the way to go when you want to let users select from a list of choices. An example is a reply in the Mail application, where an action sheet lets you choose between Reply and Forward.

As with alert views, you first need to adopt the *UIActionSheetDelegate* in your view controller's *@interface*. Then you construct the sheet with the main view controller as a delegate along with title and button strings:

```
UIActionSheet *actionSheet = [[UIActionSheet alloc]
    initWithTitle:NSLocalizedString(@"DiscoActionTitle", nil)
        delegate:self
        cancelButtonTitle:NSLocalizedString(@"Cancel", nil)
        destructiveButtonTitle:nil
        otherButtonTitles:NSLocalizedString(@"DiscoActionButton
            Emergency", nil), NSLocalizedString(@"DiscoActionButton
            Alert", nil), NSLocalizedString(@"DiscoActionButton
            Hues", nil), NSLocalizedString(@"DiscoActionButtonMixed",
```

```
              nil), nil];
    [actionSheet setActionSheetStyle:UIActionSheetStyleBlackTranslucent];
    [actionSheet showInView:self.view];
    [actionSheet release];
```

The delegate for the action sheet handles the button presses the same way the alert view does:

```
- (void)actionSheet:(UIActionSheet *)actionSheet
    clickedButtonAtIndex:(NSInteger)buttonIndex
{
    ...
    switch (buttonIndex) {
        ...
    }
    ...
}
```

SettingsViewController

The settings view controller maintains very little state: It uses only a single timer to show a preview of the flasher. And since the state of all settings is maintained in *FlashlightModel*, you don't need any additional ones.

You'll notice quite a few more views. This is typical of a settings controller that has a lot of adjustments (and associated actions) that can be performed by the user.

Coming to life

The *SettingsViewController* comes into existence because the Settings button in the main view's toolbar is hooked up to the *–showSettings* action. That method allocates and instantiates a controller instance much like *Flashlight_ProAppDelegate* did:

```
SettingsViewController *controller = [[SettingsViewController alloc]
    initWithNibName:@"SettingsView" bundle:nil];
```

But then it starts doing new things.

Sharing a model

The first difference is that it passes its model to the new controller:

```
controller.flashlightModel = self.flashlightModel;
```

This code shares the flashlight model between both controllers. That way, if one updates the model, the other will be able to access those changes without any additional work. You also get a slight savings in memory because you don't have to duplicate the instance data. With a more complex model, these savings can be significant.

Tip: Of course in some cases you want to have independent models for each view controller. In those cases, you could instantiate a new model in the controller's *–initWithNib:bundle:* method.

Flipping out and in

The first view that appeared onscreen was added as a subview to the window (page 153). You could have done something similar here by writing a lot of code to animate the views, swapping them in the hierarchy, and switching controllers. As usual, the "writing a lot of code" is your tipoff that Cocoa Touch has a lazy way to do it!

A *UIViewController,* like the one that's used as a basis for this class (page 195), can present another view "on top" of itself. The original controller is suspended while a new one takes its place. This is called a *modal view,* because the user isn't able to interact with the original while it's in this mode.

You can specify the transition into this mode with the *modalTransitionStyle* property. For settings, a horizontal flip is used, but you can also specify that the view appears from the bottom of the screen or fades in using a cross-dissolve effect:

```
controller.modalTransitionStyle = UIModalTransitionStyleFlipHorizontal;
```

Once you've set the transition, it's just a matter of calling the *–presentModalViewController:animated:* method, and the new controller and view appear with animation:

```
[self presentModalViewController:controller animated:YES];
```

From this point on, the new view controller is managed by the *UIViewController* superclass, so you can release the instance you just created:

```
[controller release];
```

Note: The other type of navigation that's common in iPhone applications is based on *UINavigationController.* This class, which is based on *UIViewControlle,* lets you "push" and "pop" new views. That way, you can easily create interfaces like the iPhone Mail or Settings app, where new views slide into place with a Back button in the upper-left corner.

The class provides *–pushViewController:animated:* and *–popViewControllerAnimated:* methods that change the user's view onscreen. There are also properties to access the navigation bar (to add your own buttons) and to change its state. The documentation for *UINavigationController* contains more information, and several code samples can be downloaded from Apple's developer site.

Picking colors. Since the settings include a picker that lets the users choose their favorite color, this view controller adopts the *UIPickerViewDelegate* and *UIPickerViewDataSource* in its *@interface.*

The data source methods *–numberOfComponentsInPickerView* and *–pickerView:numberOfRowsInComponents:* specify a single column (component) with a number of rows equal to the number of colors in the flashlight model.

Two delegate methods are used to configure the label that's displayed for each row. The *–pickerView:titleForRow:forComponent:* method provides the text for the row using the flashlight model's class method *+nameForColorIndex:*. The *–pickerView:widthForComponent:* uses the same method on each color name to determine the maximum width for the column. Remember that *+nameForColorIndex:* returns a localized string, so this is a good example of why you should never hard-code dimensions of strings. You may think 20 pixels is plenty to display "tan", but a German user will complain when "hellbraun" doesn't fit in that same space.

When the user selects a new row in the picker, the *–pickerView:didSelectRow:inComponent:* method is invoked. The implementation stops the flasher, updates the flashlight model, and then updates the views.

Interface Builder can't do it all

You may think Interface Builder has an overwhelming number of options, and it does. But even with this bounty of settings, you still can't set some things directly in the NIB file. In this view controller's *–viewDidLoad* method, you see an example:

```
[self.brightnessSlider setMaximumTrackImage:stretchableSliderMaximum
    forState:UIControlStateNormal];

[self.brightnessSlider setMinimumTrackImage:stretchableSliderMinimum
    forState:UIControlStateNormal];

[self.brightnessSlider setThumbImage:sliderThumb
    forState:UIControlStateNormal];
```

Interface Builder doesn't provide any mechanism to set the custom tracking and thumb image for the slider, so you have to do it in code. The *–viewDidLoad* method is a great place to do this, since this message is guaranteed to happen each time the view is read into memory.

Stretchable images

You might have been a bit confused with how the images for the sliders tracking images were loaded:

```
UIImage *stretchableSliderMaximum = [[UIImage imageNamed:@"SliderMaximum.
    png"] stretchableImageWithLeftCapWidth:2.0f topCapHeight:0.0f];
```

You've already seen the *–imageNamed:* method being used to load an image from your application's resources (page 182). But what's this *–stretchableImageWithLeftCapWidth:topCapHeight:* method? Are Gumby and Pokey somehow involved?

Nope—stretchable images are used frequently throughout UIKit. You can use them for buttons, gradients, and any other screen element where the size c/*hanges based upon layout changes. As you saw on page 162, view sizes can change rather dramatically as the user interacts with the app, so an image whose width and height can adjust automatically is very helpful.

When you define a stretchable image, you define two points in the original image that can be repeated when it's drawn. In the example on page 180, the second pixel from the left is specified using 2.0f. If you don't want an image to be stretched in one dimension, you can specify 0.0f as the cap: This is done for the slider's height image.

Localization

The *–viewDidLoad* method also contains an important line of code for your customers who don't speak English:

```
self.aboutBarButtonItem.title = NSLocalizedString(@"About", nil);
```

One of the design goals for the settings screen was to use as few words as possible. The icons on either side of the slider work in any language as a way to indicate brightness levels and flashing speeds. The model localizes the color names, so you don't need to worry about them, either.

That leaves one piece of text in the NIB that you need to address: the About button in the upper-left corner.

Since you have only one item to localize, it's easiest to just update the one button. When you have many items to localize, you can use another approach to localization that will be discussed shortly.

Note: The Done button is updated for you automatically since it's one of the system buttons. All you need to provide is a translation for @*"Done"* in your localized strings.

AboutViewController

A view controller doesn't need to have a model. Most do, but it's certainly not a requirement. The credits view is an example of when you don't need one. This controller also takes a different approach to doing animation.

Which version?

It's conceivable that you could have created a data model for the product's version number, but with only three lines of code in *–viewDidLoad*, the effort isn't worth it:

```
NSString *bundlePath = [[NSBundle mainBundle] bundlePath];
NSDictionary *infoDictionary = [NSDictionary dictionaryWithContentsOfFile:
    [NSString stringWithFormat:@"%@/Info.plist", bundlePath]];
self.versionLabel.text = [NSString stringWithFormat:NSLocalizedString
    (@"Version", nil), [infoDictionary objectForKey:@"CFBundleVersion"]];
```

This code locates and loads the application's Info.plist file, reads the *CFBundle-Version* property from the list, and then updates the *versionLabel*'s text by using a localized string.

The App Store does a great job of keeping your customers' software up to date. In fact, it's done so well, most customers don't even need to know which version of an app is installed. So why go to the trouble of doing this, even if it's only three lines of code?

Because you need this information when you're testing the app. There's no way to find out which version is installed on an iPhone, so if one of your beta testers finds a bug, you both need a way to identify which version caused the problem.

Note: Of course, this also means that you need to update the version number before sending out a new beta release. That's easy to do by editing the value displayed in the Info.plist with Xcode.

A different approach to animation

This controller class also demonstrates another approach to animation. Previously, you've seen views that animate themselves: *LightView* and *IFInfoView* modify their own properties by using the *self* variable. Another approach is to have the controller drive the animation of the view.

If the controller has an instance variable for the view, it can change the frame, alpha, or any other property that can be animated. Since the controller contains the logic to make these changes, this can often be a simpler solution.

Another difference with the animation performed by this class is that it interacts directly with the Core Animation framework. Using *UIView*'s +*beginAnimations: context:* and +*commitAnimations* is usually the easiest way to achieve your goals, but you'll encounter situations where you want to do something more complex.

If you look at the –*animateSpotlight* method, you see an example of this type of direct animation. In this example, it's used to move the spotlight around in the view, stopping at points along the way.

First you create the animation and define some of its basic properties:

```
CAKeyframeAnimation *animation = [CAKeyframeAnimation
    animationWithKeyPath:@"position"];
animation.duration = 15.0f;
animation.autoreverses = NO;
animation.fillMode = kCAFillModeForwards;
animation.removedOnCompletion = NO;
animation.repeatCount = 9999.0f;
```

This animation will modify the position of the spotlight view, so that property is specified as the *key path*. The duration of the animation is 15 seconds, and the large repeat count ensures that the animation will play over and over.

The next step is to add points for each keyframe. The points tell Core Animation where to move. Since position is defined as a point, a *NSValue* for each location is added to an array:

```
[values addObject:[NSValue valueWithCGPoint:CGPointMake(-50.0f, -50.0f)]];
```

A corresponding timing function is added to another array. The animation mechanism will use this function to interpolate between location values:

```
[timingFunctions addObject:[CAMediaTimingFunction
    functionWithName:kCAMediaTimingFunctionEaseInEaseOut]];
```

These arrays are then added to the animation:

```
animation.values = values;
animation.timingFunctions = timingFunctions;
```

Every view exposes a property called *layer*. This is an instance of *CALayer*—a fundamental component of the animation system. When a view is drawn, the pixels are placed in a layer that is then composited along with other layers to produce the image you see onscreen.

When you animate layers, the pixels don't need to be redrawn. And since the iPhone has a small graphics processor, the compositing operations are quick. That's the secret to how the device provides such smooth and sophisticated graphics.

To get the view's layer to start moving, all you have to do is add the animation to it:

```
[spotlightImageView.layer addAnimation:animation
    forKey:@"scrollAnimation"];
```

The key parameter defines a unique identifier that can be used to retrieve the animation with *–animationForKey:* or to remove it entirely with *–removeAnimationForKey:*.

When you're using these more sophisticated animations, you need to use the *QuartzCore* framework. This class imports the framework with:

```
#import <QuartzCore/QuartzCore.h>
```

And you'll see *QuartzCore.framework* listed in the Frameworks group in the Xcode project window.

Localized Languages: *Capisce?*

As you've been running the flashlight, you probably haven't noticed an important feature for users in Italy: The entire app has been translated into their *madrelingua* (mother tongue). Time to look at how this localization was performed.

Localizable.strings

Most of the effort in localization involves changing words. You say "white;" Italians say "bianco." In localized apps like this one, Cocoa helps you pick the right word with a file called Localizable.strings. This file acts as a small database that maps keys to values. If you look up a string with the key *White*, it will return the value *Bianco* on phones that are configured to use Italian. If any other language is selected, "White" is returned.

Note: The default language is defined as English in the Info.plist file.

Actually, it's a lie to say there's one Localizable.strings file—multiple copies exist, one for each language. If you open the Flashlight Pro project in the Finder, you see English.lproj and Italian.lproj in the Resources→Localizations folder. These folders contain the files used in each language.

When you used *NSLocalizedString()*, you performed the lookup (page 194). For example, in the *FlashlightModel*, this code is used:

```
NSLocalizedString(@"White", nil)
```

Italians will be loading files from Italian.lproj, so this line in Localizable.strings will come up as a match:

```
"White" = "Bianco";
```

Other users will have loaded the same file from English.lproj, so this will be the match:

```
"White" = "White";
```

When you're doing localization, make sure that words don't creep into other parts of your application. A great example is images: If the source text for the graphic is buried in a Photoshop layer, you'll find that your graphic designer isn't a very good translator and that your translator isn't a very good artist. Keep the text separate and your life will be much easier.

Localization logistics

You have two major hurdles when doing localization: finding a native speaker to do the translation and testing the final product. As you look through the Italian words in Localizable.strings, it looks great. There are plenty of accents and cool-looking foreign words. There's just one problem: A native speaker did not do this localization.

Note: Your humble author spent several years living in Italy. This language perfect he has not, but good it is.

Don't use automated tools or get your second cousin's best friend who spent 6 months overseas to do the translation. A native speaker will be able to tell that it's wrong and you'll make a bad impression. For many users, a bad localization is harder to use than the original English.

Once you've done the translation, you encounter the second problem. Unless you're an amazing polyglot, there's no way you can know if the localization is correct. You need to find someone who speaks the language and to get him or her to test it for accuracy.

You should, however, verify that the localization is functionally correct. You want to make sure that all the words have matches during the lookup phase.

The easiest way to do this is by using the Settings application in the Simulator. Tap the settings icon, and then pick the first item in the list (General). In the next menu, tap the third item from the top (International). Then pick the first item in the list (Language) to see a list of all available language settings. (Knowing the relative position of all these buttons is important: You need to know which buttons to push in order to switch back to English!)

While you're changing your language settings, try setting Italian and running the Safety Light application. Seeing the localization in action will give you a better feel for how it all works.

Note: Why go through this hassle? About half of your revenue in iTunes will come from outside the United States. The more you cater to these users, the more you sell. No one ever promised that making money would be easy.

Layout breakage

Sometimes differences in language cause view layouts to change. For example, the button width for "Tap to close" in English is going to be very different than its literal counterpart in German: "Zum Schließen antippen".

One approach your translator could take is to find a similar meaning that's shorter. For example, "Tap to close" could be shortened to the German word for close: "Schließen". The width of both phrases is similar, making the visual design easier without sacrificing comprehension.

Note: Apple uses this technique with some of its system controls. For example, "Entriegeln" is displayed instead of "slide to unlock" when you press the Power button on the device.

Another approach to handling phrases with different lengths is to make views adapt to different sizes. The *IFInfoView* that drops down from the top of the screen is an example. It adjusts its height based upon the message you want to display. If it encounters a long phrase in German, it just displays multiple lines.

A final technique to aid in localization is to use symbols whenever possible. If you've ever traveled abroad, you know that it's much easier to encounter restroom doors with pictures of men and women rather than signs saying "Damen" and "Herren." Hope you guessed right!

This approach was taken with the settings view. The brightness and flasher speed use icons instead of words. No translation is required and users in all languages can understand the function of the sliders.

AboutView.xib

Despite all your efforts to just translate the words, sometimes localizations require a separate NIB file. For example, if you're doing a help screen for a game, you're going to have a lot of text that flows around images and controls. It's easier to change the layout for a whole screen than it is to adapt the text to a single layout.

The AboutView.xib file is an example. You'll see that the file appears in both the English.lproj and Italian.lproj folders. Cocoa loads the correct NIB based upon the user's language settings. You can change any part of the layout; just make sure that the actions and outlets are maintained. If the connection between a button and its action is broken, you have a bug that only users of that language will encounter.

GEM IN THE ROUGH

How to Localize a File

You may have noticed that localizable files show a disclosure triangle in the Groups & Files list. When you click the control, a list of languages shows. Xcode treats localizable files differently than normal files.

Here's how to create a localizable file:

1. In Xcode's Groups & Files list, select the file you want to localize.

2. Choose File→Get Info to bring up the File Inspector window.

3. In the File Inspector, click the Make File Localizable button. This option moves the existing file into an English.lproj folder. At this point the disclosure triangle also appears.

4. In the File Inspector, click the General tab, and an Add Localization button appears in the lower-left corner. Click it.

5. You'll be prompted for a new name for the localization. You can enter *Italian, French, German, Spanish, Japanese,* or any other language the iPhone OS supports.

Wrapping It Up

This iPhone app is looking good. It's now time to start beta testing and getting it ready for sale on iTunes. Your coding phase is about to end, but you still have a lot of work to do!

Part Three:
The Business End

III

Finishing Touches

You've now finished coding the best thing ever seen in the App Store. Or have you?

Now's the time to put the final spit and polish on your app. That includes testing your creation with a wider group of people and their devices. Since your development is winding up, now's the perfect opportunity to clean up your interface graphics. Then it's time to start working on screenshots and other promotional materials for a product website.

Beta Testing

Testing your application before submitting it to the App Store is an important part of the development process. Getting your work in front of other people will help you find problems with both your design and code.

Your App Works for You...But

You and your development team already know how the application works. You're intimately familiar with how each component should function. The code runs perfectly on your test devices.

In other words, you're the exact opposite of every single person who will download your product from the App Store.

Users find the darnedest things

Giving your application to people who've never seen it generates a lot of great feedback. As you learned on page 113, customers are your greatest feedback mechanism, and beta testers are your first customers.

When you give your application to other people, two things will happen:

- They'll find bugs in your code.

- They'll find things they don't understand.

Finding bugs in your brand-new code may be disappointing, but it's much better than the alternative. If you release something to the App Store and find that a large percentage of your users are having an issue, you'll be in quite a predicament. It takes many days for a new version of the app to be reviewed by Apple and made available for users. During that time, customers will be leaving bad reviews in iTunes, you'll get bad press, and you'll be sweating bullets.

But if a large portion of your *test* group doesn't understand how a feature works or gets frustrated using your app, that's great news. It means one of your design decisions was wrong. Don't get upset when you hear this feedback: Every developer has gotten such feedback at one point or another. Now's your chance to put things right before customers in the App Store complain about the same thing!

Flashlight improvements

In fact, the flashlight you saw in the last chapter went through some design revisions based on tester feedback. You may encounter some issues like these in your own apps:

- Several people in the test group found that it was cumbersome to change the brightness quickly. They had to go in and out of settings each time to adjust the light level. What they really wanted was a simple gesture to lower the brightness temporarily. That's how the single tap in *LightView* was born.

- Another tester complained of how easy it was to inadvertently turn off the flasher. She was using the Safety Light in an armband during her evening workout and accidentally tapping on the toolbar while running. The single tap gesture in *LightView* was extended to a double-tap that locked the screen.

- The designer-type in the bunch wanted some visual flair in the app. The settings view looked cool enough, but he wanted something that no other flashlight on the App Store had—batteries! The background that you see when flipping to settings was an easy feature to add by putting a *UIImageView* in the *MainWindow*.

The flashlight's code had a few bugs that you wouldn't have found without the help of the beta testers:

- The Italian localization didn't display correctly in some cases because of problems in *Localizable.strings*. This user also discovered an issue with long messages not being displayed correctly in *IFInfoView*.

- Another tester found a bug that had never been considered by the development team: You could dim the brightness of the light and start the SOS flasher. That prevented the warning signal from being as effective as possible.

Luckily, there weren't any serious problems, and the beta test went smoothly!

Ad Hoc & Roll

So how do you get this beta version to your testers? With a mechanism called *Ad hoc distribution*. This technique lets you share your iPhone app with up to 100 testers. But first you need to make a few more changes to your project so these users can run it on their iPhone or iPod touch.

Gather your testers

The first step is to collect device IDs from potential testers. You can use the same techniques as you used on page 137. For most of your testers, clicking the serial number in iTunes or running the free Ad Hoc Helper application from the App Store is the easiest way to get the 40-character UDID string.

As a part of this process, also get the testers' email address and the type of device they have. You'll use the email addresses to set up a mailing list. Knowing which type of device they're testing on will be helpful in debugging hardware-specific problems.

Tip: Remember that your iPhone developer account limits you to 100 devices for the entire year. Keep the number of people in your beta test well below this number.

Be aware that adding each new tester increases the amount of work it takes to run the beta test: You'll be responding to additional feedback, helping with more installation problems, and doing additional tester support. For most applications, you'll get plenty of test coverage with a group of 20 to 40 people.

Distribution

Up until now, you've only been provisioning devices for development. These profiles are designed to work within the Xcode environment. For example, a development profile includes a configuration value that allows the debugger to attach to your application's process after it's launched.

Your testers won't be using Xcode, so they're going to need a different kind of profile: one for general distribution. Luckily the process for creating these profiles is very similar to the one you've been using for development, so no surprises should occur.

Certificates. As with your development profile, the first step is to generate a certificate using Keychain Access:

1. Use the Keychain Access application's assistant to generate the request. To start, select Keychain Access→Certificate Assistant→Request a Certificate From a Certificate Authority.

   The Certificate Assistant dialog box opens.

2. **In the User Email Address field, type your email address.**

   Use the same one you used when signing up for the iPhone Developer Program.

3. **In the Common Name field, type your (or your company's) name. Turn on "Save to disk" so a file is created.**

   If you signed up as an individual developer, use the same name you used to enroll as an iPhone developer. If you signed up with a company account, use the same company name.

4. **Click Continue.**

   You'll be prompted to save the certificate request.

5. **Change the name of the file to *Distribution.certSigningRequest,* and save it on the desktop.**

6. **Click Done to close the assistant after you see the confirmation screen.**

Then it's time to upload that certificate request to the Provisioning Portal. (If you need a refresher on what the portal is and how to get there, see page 136.)

1. **On the Provisioning Portal page, click Certificates in the left column.**

   The Certificates page opens to the Development tab.

2. **You're going to create a distribution certificate, so switch to the Distribution tab. You see a warning that you don't have a valid certificate; click the Request Certificate button so you can upload your request.**

   Click Choose File, and a file dialog box opens (Figure 7-1).

3. **Navigate to the *Distribution.certSigningRequest* on your desktop and then click Choose.**

4. **Back on the Distribution tab, in the lower-right corner, click Submit.**

   You see a confirmation that the request was submitted. The certificate's status reads "Pending Approval". Believe it or not, Apple is asking you to approve your own request.

Note: If you're working with a company account, only the team agent can approve the certificate request. The agent is the person who originally signed up for the developer account.

---

5. **In the Action column, click Approve.**

   The status column changes to Pending Issuance.

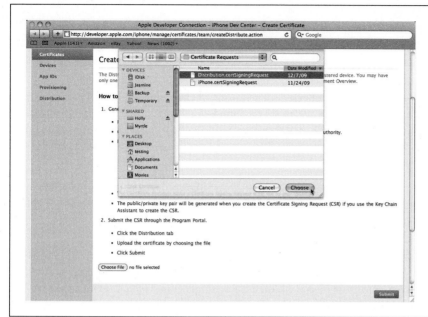

Figure 7-1:
You request a distribution certificate on this Provisioning Portal page. Simply upload your certificate signing request, submit it, and approve the request.

6. **After a short wait, you should be able to refresh your browser and see a Download button; click it.**

   When the download is done, you should see a *distribution_identity.cer* file in your Downloads folder.

7. **Double-click to open the *distribution_identity.cer* file.**

   You're prompted to confirm the addition of this new certificate.

8. **Make sure the Keychain pop-up menu is set to "login" and then click Add.**

   At this point, your keychain has been set up correctly. You should see iPhone Distribution along with your name in Keychain Access.

As with your developer certificate-signing request, it's a good idea to save your distribution request file for later use. It's one less thing to hassle with when your beta build is broken because of an expired certificate.

If you're using a company account, each of the developers on the team has an individual code-signing certificate for development. With a distribution certificate, you have only one for your entire company. That's because Apple wants to verify which company created the software, not the individual developer who did the build.

Tip: If you ever need to check the contents of a certificate-signing request, use this command in a Terminal window to view it:

```
$ openssl req -noout -text -in Distribution.certSigningRequest
```

The output will show the email address and common name (CN) used for the request.

Devices. Once you've collected the relevant information from your beta testers, head over to the Provisioning Portal, and spend some quality time in the Devices section. Adding devices for testers is exactly the same process you used on page 144. The portal doesn't differentiate between devices that are used for development and distribution.

Make sure to use the feature that lets you add multiple devices with a single click of Submit. Click the + icon at the end of each line to get another input field for the device name and ID; it makes this tedious job a little bit quicker.

App IDs. For this example, you're distributing a beta application from your company. You want your App ID to reflect that. Say that your company is Squidger Industries, and your Internet domain is squidger.com. Here's how to create an App ID called Squidger Apps:

1. **On the Provisioning Portal page, click App IDs in the left column.**

   You should see the App ID named All Apps that you're using for development.

2. **Click New App ID to add the new ID you're going to use for distribution.**

3. **For Description, type *Squidger Apps* as a reminder that you're using this App ID for building and running Squidger apps. Leave the pop-up menu set to Generate New.**

   Next enter the pattern used to match the applications.

4. **Type *com.squidger.** for the Bundle Identifier and then click Submit.**

   After you return to the main App ID page, you see Squidger Apps in the Description column (Figure 7-2).

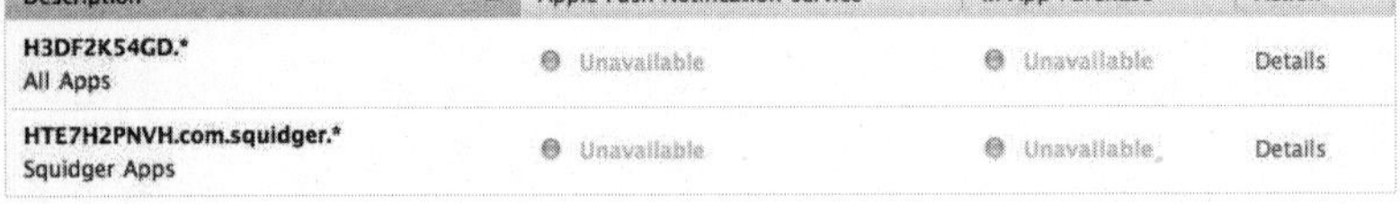

Description		Apple Push Notification service	In App Purchase	Action
H3DF2K54GD.* All Apps	▲	⊖ Unavailable	⊖ Unavailable	Details
HTE7H2PNVH.com.squidger.* Squidger Apps		⊖ Unavailable	⊖ Unavailable	Details

Figure 7-2:
After you create the App ID for Squidger Industries, Squidger Apps is listed in the portal along with All Apps. This new App ID is used for distribution, in a way similar to the one used for development.

You're now ready to pull it all together in the provisioning profile!

Provisioning. As you saw with certificates, there's an additional tab when you select the Provisioning section in the Provisioning Portal. Here's where you create and manage the profiles used for distribution. For now, you're just going to create one for your beta test, but eventually you'll be adding one to distribute your app on iTunes.

After switching to the Distribution tab, click New Profile, and you see a screen that looks very much like the one you saw for development. The big difference is that you now see a Distribution Method listed. Since you're not quite ready for the App Store, click Ad Hoc and you'll be rockin'.

As you fill out the Profile Name, think for a moment about your testers. They like to help developers, and it's probable that they're also testing someone else's software. If you use the name "Beta Test" for the profile, that's going to be confusing when it shows up on the testers' phone. And when their Beta Test profiles expire, they're not going to know whom to ask for a new one.

It's best to use your company name for the Profile Name. If you enter Squidger Beta Distribution, for example, your testers will know it's the profile for testing Tiddlywinks Pro. This descriptive name will also help you, as you'll see shortly.

For the App ID, select Squidger Apps, so your testers can run any and all applications from the brilliant minds at Squidger Industries.

Note: If you're using push notifications or doing in-app purchases, you need to select the App ID that you've configured with these features. Both these features require an identifier that does *not* use the wildcard (*) character.

Under Devices, click Select All, and all your beta testers will be added to the distribution provisioning profile.

Tip: Remember, if you add a new device for a beta tester, you need to update this profile and get the tester a copy of the *.mobileprovision* file. It's *very* easy to forget this important step, and it causes confusion for both you and the tester.

Click Submit, and the portal starts generating your distribution profile. You should be able to refresh your browser in a few seconds and see a Download link. Click that link and then add the Squidger_Beta_Distribution.mobileprovision file to Xcode by dragging it onto the application icon.

Save the *.mobileprovision* file in a safe place. You're going to be sending it to your beta testers.

Entitlements

Now that you've gotten the provisioning profile sorted out, shift your attention to Xcode's build settings.

The primary difference between signing code for development and for a distribution profile is the configuration parameter that allows Xcode to debug your process. The default value for the *get-task-allow* key enables debugging, but when you're doing Ad Hoc distribution, this setting must be turned off.

To accomplish that, change the Entitlements configuration used during code signing. Here are the steps:

1. **In Xcode, choose File→New File. Then, in the New File dialog box, choose Code Signing in the iPhone OS source list.**

   You see two templates, Entitlements and Resource Rules, both of which let you configure code signing.

2. **Select the Entitlements template and then click Next.**

3. **Set the File Name field to *dist.plist* and Location to the top level of the project folder. Make sure that the file is added to the correct targets. Click Finish.**

   The file should now appear in the property list editor.

4. **Turn off the *get-task-allow* checkbox and then save the file.**

You now have a file that can disable the debugger. All that's left to do is to add that file to a target.

Tip: Don't be afraid to be lazy and to use the *dist.plist* file that's in the Flashlight Pro application. The steps above have already been done for you!

As mentioned on page 127, it's convenient to put your beta release in a separate target. Since target settings override their counterparts in the project and Xcode, it's the perfect place to define how you want the beta built.

In fact, this was done in the Flashlight Pro project. If you open Targets, you see a target named "Safety Light (Beta)". When you double-click this target, the window that contains the build settings appears. Make sure Configuration is set to Release, and scroll down to the Code Signing section; you see the Code Signing Entitlements set to *dist.plist*.

Targets or Configurations?

Some developers create a Distribution build configuration as a place to store beta settings. Instead of selecting a new build target, they just change the Active Configuration before doing their build.

The problem with this approach is that you now have two "release" configurations. If you change one of the build settings in the Release configuration and forget to update the same one in Distribution, your beta testers aren't going to be working with the same code you are. You also end up with many configurations as you add the final release, a Lite version of the product, a Lite version of the final release, and so on.

When using multiple targets, you have a similar problem: You can add a file to one target and forget to add it to the other targets. In practice, it's fairly easy to catch this error—you see different file counts when you use the disclosure triangle in Groups & Files. All targets should show the same number of items in parenthesis after Copy Bundle Resources, Compile Sources, and Link Binary With Libraries. Also, with a missing file in a target, you're likely to see a compiler error that alerts you about the discrepancy.

Another advantage to using targets is that they can be used to automate your build process. You can create "aggregate" targets that let you run scripts that perform postprocessing. To create an aggregate target, go to Groups & Files, right-click Targets, and then select Add Target. In the window that appears, select Other in the source list, and then choose Aggregate from the Other list.

The Flashlight Pro project uses aggregate targets for both the beta and final application builds. Under Targets in Groups & Files, you see two red target icons: Beta and Final.

If you open the disclosure triangle for Beta, you see that the aggregate target first builds Safety Light (Beta) and then runs a script named Package Release. If you double-click the script name, you can edit it. This script makes a backup of the dSYM debugging information (for processing crash reports) and creates a ZIP file that you can distribute to testers.

Several scripts can run sequentially. In the Final target, the release is packaged as it was during the beta build and then resigned so that you can run it on a test device.

You can add your own scripts to an aggregate target by right-clicking the target name and choosing Add→New Build Phase→New Run Script Build Phase. A good example would be a script that uploads your beta build to a server so your testers can download it.

To automate this process even further, the flashlight project has two shell scripts. After running *build_beta* or *build_final* from a command line in the Terminal, you get a ZIP file that's ready to distribute to beta testers or to upload to the App Store.

After you change the entitlements build setting, you also need to change the Code Signing Identity setting that's a couple of lines down. Use Any iPhone OS Device along with iPhone Distribution. The automatic profile selector should show that it matches Squidger Beta Distribution or whatever name you provided when you set up the distribution profile.

Build and send

Now that your build settings are shipshape, it's time to do the build. Before you throw the switch, make one last edit in Xcode: Change the "Bundle version" in Info.plist to something that uniquely identifies the release. If this is your first beta test for your first version, use something like "1.0b1". Now go to Project→Set Active Target, select your beta release target, and do a clean build (page 148). If all goes well, shortly you see "Succeeded"!

Note: The Flashlight Pro project has two beta targets. The first is named "Safety Light (Beta)" and it builds the application. The second is an aggregate target named "Beta" that builds the app using the first target, creates a ZIP file, and uploads the app to the download server.

After your build is complete, make a Zip file of the application that you're going to send to testers. Go to the target's build folder, and you see something like Figure 7-3.

Figure 7-3:
After a build, your target's folder has an application with a circle-backslash symbol and a dSYM file. Both the Mac and iPhone versions of OS X use an Info. plist file. Don't worry— the symbol simply indicates that the Mac and iPhone have different processors. The file with a .dSYM extension contains debugging symbols. Compress the application file before sending it to beta testers.

Tip: You can determine your target's build folder by looking at the Per-configuration Build Products Path. The default values are *build/Debug-iphoneos* and *build/Release-iphoneos* for device builds. The Flashlight Pro project prepends "Beta–" and "Final–" to the folder names so your beta and final builds don't get mixed up with your test builds.

The quickest way to make the Zip file is to Control-click (or right-click) the app icon and to select Compress from the shortcut menu. Save this Zip file in a safe place, since you'll be sending it to your beta testers. You may also want to rename the file with the same version number that you used in Info.plist, which will make it easier for you and your beta testers to track multiple releases. Something like *TiddlywinksPro-1.0b1.zip* works well.

Remember that *.mobileprovision* file you saved a few pages back? Time to dig it up to send along with the application Zip file. Then attach both the TiddlywinksPro-1.0b1.zip and Squidger_Beta_Distribution.mobileprovision files to your email to beta testers.

But before you send that email, you need to do one more test. You need to make sure that your testers will be able to install and run the software.

Install

Your beta testers probably won't have Xcode and its Organizer window to install profiles and applications. And if you have a tester who's running Windows, they certainly won't be able to run this Mac application. So what's the trick?

Actually, there are two tricks. The first, and easiest, is to use iTunes!

iTunes. The first thing your beta testers need to do after receiving your beta software is to decompress the Zip file. Just double-click the file in the Finder, and the app icon with the circle-backslash will appear. At this point, installing the app is just a matter of dragging the application and the provisioning file onto iTunes' application library (Figure 7-4).

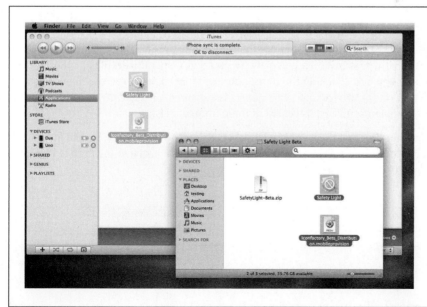

Figure 7-4:
Testers can use iTunes to install your beta application and provisioning. They can just drag the app and the .mobileprovision files into iTunes' application library.

The applications added to iTunes behave very similarly to the ones that are purchased from iTunes. The only difference is that they show "Unknown Genre" instead of the category name. You can even use File→Get Info to display the version number. Not surprisingly, you can then sync the beta application to the device (Figure 7-5).

Figure 7-5:
The beta application can be synced to the tester's device just like any other application from the App Store. Just make sure that there's a checkmark next to the application name before you start syncing.

> **Note:** Testers on Windows can get confused about what's in the Zip file. Mac OS X adds some additional information that makes it hard to find the application in the archive. Luckily, you can take additional steps to make the process simpler: *http://www.markj.net/iphone-ad-hoc-distribution-windows-mac/*.

The syncing process will also move the provisioning profile onto the device. You can check that it was installed correctly by going to General→Profiles in the iPhone's Settings app and checking the name and expiration date. If you see the wrong one, delete it on the device and sync again. Any expired or duplicate profiles will only cause problems: It's best to get rid of them immediately.

After checking that the correct profile is installed, you should be able to launch the app on the phone. Check your credits screen to make sure that the proper version is installed. (Now you see why having that view is useful!)

Note: In Figure 7-5, you see that Safety Light shows custom artwork in iTunes (the app icon has "BETA" written on it). That's accomplished by an *iTunesArtwork* file added to the beta target. The file is a 512 × 512-pixel PNG file with the extension removed.

The Flashlight Pro project shows how to do this in the Beta target. If you don't include this file, everything will work correctly, but your testers will see a generic icon. Adding the text on the icon reminds your testers that they're using a beta release.

Make sure that you don't add the *iTunesArtwork* to your final build that gets uploaded to the App Store. Apple adds this artwork after submission and if it's already there, it can cause problems.

iPhone Configuration Utility. The second trick is to use the iPhone Configuration Utility. Apple provides this tool for both Mac and Windows at *http://www.apple.com/support/iphone/enterprise/*. It was originally developed for enterprise deployments of iPhone applications, but it also works great for beta testers.

Note: Ad Hoc distribution was first introduced during Steve Jobs' WWDC keynote in 2008. It was billed as a way for enterprise customers to distribute their apps. Independent developers like you had other ideas and quickly put Ad Hoc distribution into use for beta testing.

After you download, install, and launch the application, you see a screen like Figure 7-6.

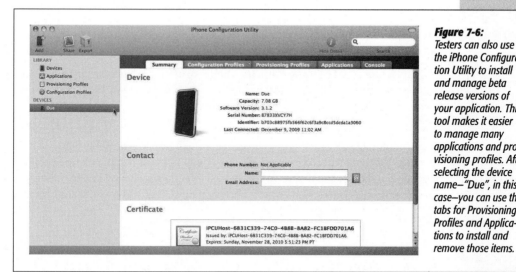

Figure 7-6:
Testers can also use the iPhone Configuration Utility to install and manage beta release versions of your application. This tool makes it easier to manage many applications and provisioning profiles. After selecting the device name—"Due", in this case—you can use the tabs for Provisioning Profiles and Applications to install and remove those items.

When using this utility, select Applications in the source list, and then drag in new applications to update the list. Similarly, you can drag your *.mobileprovision* files into the list displayed by Devices.

After registering both these items, switch over to the named device in the source list, and use the Provisioning Profile and Applications tabs to install and uninstall. Unlike with iTunes, you can also see the dates that profiles expire. Another big advantage is that you can install without synching, which is very important for someone who's testing at work but whose device is bound in an iTunes media library at home.

Again and again. Unless you're extremely lucky, you'll have more than one beta release. You need to advise your testers on how to update their existing beta install. For testers using iTunes, the process is simple. All they need to do is drag the new version of the application into iTunes. They see a warning that the application already exists, and use Replace to update to the new version.

If a new provisioning profile is needed, testers should delete the existing one on the device and then drag a new one into iTunes. Another warning occurs before the file is replaced. Once the updated files are in place, sync will get the new version onto the device.

If the testers are using the iPhone Configuration Utility to install the beta versions, they must completely uninstall the app before doing a fresh install. This wipes out any existing documents, so game scores and other application data are lost.

Note: After you add a new version of an application, the version number may be displayed incorrectly. If you want to see the correct information, you need to quit and restart the iPhone Configuration Utility.

Crash act. Your code is perfect, right? Yep, that's what every developer thinks until one of those crafty beta testers gets hold of it.

Crashes happen. And it's a good thing; it's better to catch the bad memory references, missing methods, and other bad guys before your customers in the App Store do. Every crash you see now is one you won't see later.

The iPhone OS keeps track of every crash that happens on the device. When they occur, it automatically generates a file called a *crash report*. Normally, these files are uploaded to Apple when a user syncs with iTunes. Remember that dialog box that popped up in iTunes the first time you synced after an application crash? When you saw "Your device contains diagnostic information...", that was the first time a crash report was detected and sent to Apple.

You can get these crash reports directly from your beta testers. Apple has described the process in Technical Note TN2151 (*http://developer.apple.com/iphone/library/technotes/tn2008/tn2151.html*). In short, the tester can extract the *.crash* file from iTunes on their Mac or PC and mail it to you. Then you open the Organizer window in Xcode and drag the file into the list of Crash Logs under IPHONE DEVELOPMENT. After the crash report is processed, numeric memory addresses are transformed into class and method names with line numbers that correspond to your source code.

Note: The most important part of this step is that you save a copy of your dSYM file and your application binary. It has to be the exact one you sent to your testers, and the files must be archived in the same folder. That's why the Flashlight Pro project automates the process of saving these files by date and time.

Once you have this information, it's usually easy to track down the source of the bug.

Clean Up Your Act

While you're beta testing, you shouldn't be doing active development. If you do, your testers are working with a moving target. It's in everyone's best interests if you freeze the feature set and concentrate on fixing problems that pop up.

Given this "don't touch the code" edict, beta testing is a great time to start cleaning up the appearance of your app. And if you're doing localization, it's the right time to gather translations and distribute them to the testers.

Beautification Committee

Developers write beautiful code. Designers draw beautiful graphics. Figure 7-7 shows what it can look like when a developer makes graphics.

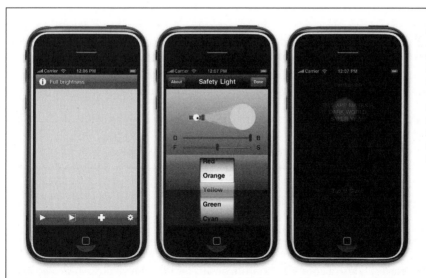

Figure 7-7:
The flashlight user interface with place-holder graphics used during development. The beta test can be a good time to clean up the visuals in your application, because it can be done without modifying much code.

And when the designer comes to the rescue, everyone benefits. Check out Figure 7-8.

Figure 7-8:
The same screens as in Figure 7-7, but with graphics produced by a designer. Notice how much nicer the toolbar icons, settings, and credits screen look when they're done by a professional? It's an exciting time in the life of the project when this work gets done!

For most developers, seeing the final graphics is a magic moment. The crappy graphics you threw into the app get replaced, and your application feels a thousand times better. It's also a special time for the designer: She finally gets to see her pixels put into motion. Everyone's vision is brought to fruition.

As a part of this cleanup process, it's often helpful to give your designer written instructions for each file. As the developer, you know how ToolbarDiscoStart.png is being used because it's right there in the source. Most designers don't read code, so give them a break; be explicit about what you need. Make sure to specify any special size, color, or placement requirements that the graphics need.

Tip: The flashlight project includes a manifest file in the Images folder that explains each graphic. This way, you can give the entire directory to the designer, and she has everything she needs.

Speaking in Tongues

In Figure 7-8, you may have noticed that the interface is being displayed in Italian. *Che bello!*

The beta test is also a good time to work on any localization you're going to do in the application. If you've been diligent about using *NSLocalizedString()* in your code, it's relatively easy work. Create a localization package for your foreign-language translators. This package includes the *Localizable.strings* file and any NIBs that need to be updated. Put everything in a Zip file and send it.

Once you get back the localized files, you can add them to your project and do a build. Make sure that your translators are a part of the beta test, because they need to check the final results. You don't want any *idiotzia* slipping through!

Web Development

Once you have your graphics updated and localization started, you can start to think about a promotional website for the product. Some developers think that the only promotion that they need is iTunes itself. For some simple products, that approach works fine. But if you have a product with any complexity, you need a way to show off the features to potential customers.

The App Store only allows you to show five screenshots. And it has no way to show a movie of your product in action. Once people can see how great your product is, they'll want to buy it. A product website gives you room to explain your product in greater detail than is possible within iTunes.

Work with both your designer and marketing person on this endeavor. It's also important to realize that it's going to take awhile to develop the visuals and sales pitch for the site. That's why you want to get an early start on this activity.

The Site

As a developer you can help everyone by putting some of the basic elements in place. For example, put together the basic structure of the site in HTML, and get the web server set up to host the files.

Note: The Missing CD (*www.missingmanuals.com/cds*) contains the HTML and CSS source code for the Safety Light website. It's designed to be adaptable for many different kinds of applications, so feel free to use it for your own products.

As you're working on this site, it's important to remember that *many* users will be visiting from their iPhone or iPod touch. You need to make sure that any design you come up with works with a small screen format. It's surprising how many developers never bother to open their product website on a device!

Content

For the most part, the iPhone can display the same kinds of images that your desktop browser can. Be careful how you present those images. Avoid JavaScript that does fancy zooming of screenshots: The zoom will be frustratingly slow, and the "full window" display is hard to navigate on the device without pinching.

On this website, one line of additional HTML code is also very important:

```
<meta name="viewport" content="width=999" />
```

This line gives Safari a hint about the width of your site so that it can optimize the display. Replace "999" with the actual width you want to be zoomed to when your site is loaded. You can get the scoop on these and many other tricks that web developers can use with the mobile version of Safari in these articles at A List Apart:

http://www.alistapart.com/articles/putyourcontentinmypocket/

http://www.alistapart.com/articles/putyourcontentinmypocketpart2/

Spend a few minutes familiarizing yourself with the capabilities of Mobile Safari—your future customers will love you for it.

Movies

If you're going to include a movie on your product website, you *must* follow one rule: no Flash. Safari on the iPhone can't display it, so all the work that went into making a great pitch for your product will have been for naught.

Instead, use QuickTime to create the movie. This application, shown in Figure 7-9, can take a single movie and optimize it for display on the desktop and on mobile devices.

The export process also creates some HTML and JavaScript that makes it easy to incorporate the output into your website. The output folder that you specify contains a *ReadMe.html* file (Figure 7-10) with the details.

Another possibility for movies is to use YouTube. Yep, even though YouTube uses Flash, you're not breaking the rule. YouTube converts the movies you upload into different file formats. When you view the movie on the desktop, it's displayed in a Flash player. On the iPhone, H.264 video is used—the same format that's used in the iPhone's YouTube application. You won't have as much control over the presentation, but YouTube can be a cheap and effective way to promote your product.

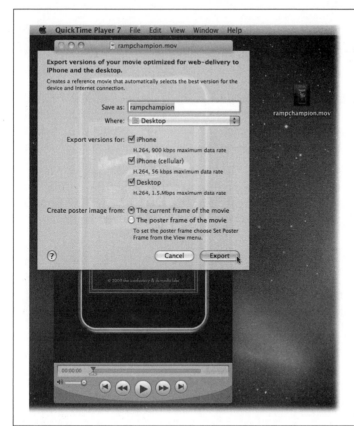

Figure 7-9:
The movie rampchampion.mov *is being converted into formats that are compatible with the desktop and iPhone (both via wireless and cellular networks). The resulting files are placed in a folder called* rampchampion *on your Mac's desktop.*

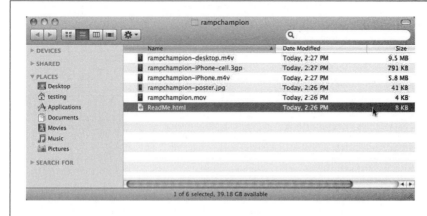

Figure 7-10:
The exported files in the rampchampion folder. Notice how the desktop file is 9.5 MB, while the file for the iPhone on a cellular network is only 791 KB. Open the ReadMe.html file in your browser, and you see code samples that show you how to use the movies on your website.

Tracking

As customers start to visit your product website, you'll want to know how many there are and where they're coming from. This information will lead you to websites that review your application, people who mention it on Twitter, and online forum comments. You'll get a very good idea of what people are saying about your work.

The best way to track visitors to your site is with a product called Mint *http://haveamint.com*. This simple system records all visitors to the site by adding a single line of JavaScript code to your HTML header. You need PHP and MySQL to install this product, but most web hosting companies provide this free of charge.

To get an idea of how it works, you can take a look at the Safety Light application's website statistics *http://safetylightapp.com/mint/*. Don't worry about your competitors being able to do this on your website: The standard Mint configuration uses a password to protect this information. On this site, it's been enabled just to demonstrate the capabilities.

App Store, Here You Come

While your beta test is ongoing, it's also the perfect time to get yourself connected to iTunes. While testers are putting the product through its paces, you can get started on the work needed to get it ready for sale!

Time to visit another Apple website for iPhone developers....

For Sale

Y our beta test is going well, so now it's time to turn your attention to how you're going to sell the final product. iTunes has over 100,000 applications, and it's time for yours to join them.

This chapter will introduce you to iTunes Connect, a web application that provides access to the database of products offered in the App Store. You'll use your browser to set up financial details, provide information about your application, set or update pricing, and to upload the final build.

You'll also encounter the review process that Apple uses for every submission. It's one of the more frustrating areas of iPhone development, so this chapter will show you how to smooth out the process.

Sign on the Dotted Line

The first thing you do in iTunes Connect is deal with the legal contracts and banking information with your new partner.

Welcome

To log into iTunes Connect, point your browser at *http://itunesconnect.apple.com*. Take a moment to bookmark that URL—you'll use it a lot. On the login screen, enter your iPhone Developer Center username and password.

Tip: If you're already logged into the developer site, you can simply click the iTunes Connect link in the right column on the main page. You won't need to log in again.

After logging in, you see a form that asks you to accept the terms and conditions for the website. You only see this the first time you use iTunes Connect or if Apple changes the contract. If you want to put your app in iTunes, accept these conditions.

Once in iTunes Connect, you see a screen that looks something like Figure 8-1.

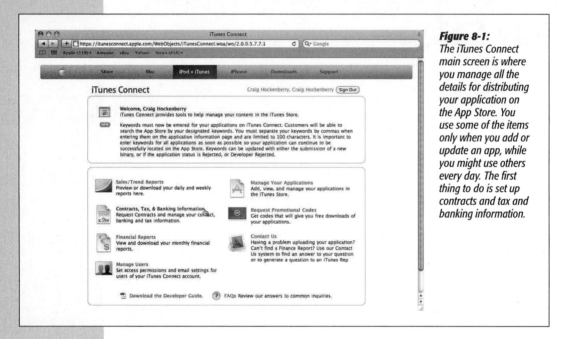

Figure 8-1:
The iTunes Connect main screen is where you manage all the details for distributing your application on the App Store. You use some of the items only when you add or update an app, while you might use others every day. The first thing to do is set up contracts and tax and banking information.

First Order of Business

The first order of business is to get your contracts, tax, and banking information set up. When you signed up for your iPhone Developer account, you provided only contact information. But a business partnership involves a lot more than knowing each party's identity. It's time to get the lawyers, government, and financial institutions involved!

Start getting things in place by clicking the "Contracts, Tax, & Banking Information" link on the iTunes Connect home page. Once this page loads, you see that you already have a contract in effect. When you signed up to be an iPhone developer, Apple gave you a contract to distribute free applications worldwide. If you don't want to make any money from iTunes, you're done!

Contracts

As with most things involving money, things are a little bit more complicated if you want Apple to pay you. Turn on the Request Contract checkbox and click Submit. (If someone else on your development team is handling the legal stuff, that person should complete these steps.)

Once you submit the request, you see the contract that Apple's offering you—Paid Applications. The contract assigns Apple as your agent/commissionaire to sell and distribute your software with no ownership interest. You agree to provide licensed applications and to support the end user. The contract also stipulates Apple's commission (30 percent), how taxes are collected worldwide, and when you get paid (net 45 days from the close of the month).

Note: As with any contract, you should have both your lawyer and accountant review the entire document. The previous paragraph is just a summary.

If you and your lawyer accept the conditions, click the Agree checkbox and submit the form. You see a confirmation screen, and a PDF of the contract is on its way to you by email. Click Done to return to the contract setup screen.

Contact info

The next step to set up the contract is to add additional contact information. iTunes Connect lets you assign specific roles to various people in your organization. For example, your accounting department can get access to the financial reports, while marketing can generate promotion codes.

1. **In the Contact Info for your contract, click Edit.**

   The next screen shows your company's legal address along with the roles within iTunes Connect. For now, just assign yourself to all the roles. As needs arise, you can create new users, with new Apple ID accounts, by clicking Manage Users on the home screen.

2. **To create your contact information, click Create New Person and then enter your iTunes Connect user name and email address. Add your title and phone number and then click Create.**

   You end up back on the contact information page.

3. **For the Senior Management, Finance, Technical, Legal, and Promotions roles, choose your name from the drop-down menu.**

   Once you're done, click the Submit button to proceed to the next step.

Bank info

After your contacts are defined, you need to tell Apple where to deposit the money you earn each month. The money will be deposited in two ways:

- **ACH.** A standard electronic deposit using the Automated Clearing House. This method is used only for earnings in U.S. dollars. Deposits from "Apple Inc." are for proceeds in the U.S. iTunes store. You'll also see an "iTunes S.A.R.L." deposit for countries without their own iTunes store.

- **SWIFT.** A wire transfer using the Society for Worldwide Interbank Financial Telecommunication. Earnings from the iTunes stores in Australia, Canada, Europe, Great Britain, and Japan are deposited using this mechanism. This transfer also converts any foreign currencies into U.S. dollars using the current exchange rate. Unfortunately, many banks don't provide information about the source of the transaction or the rate used, making it difficult to reconcile your records.

Both transfer mechanisms need an account number in order to complete the deposit. An ACH transfer uses a nine-digit ABA routing number to determine where the money will be deposited. A SWIFT code is used for the wire transfer between foreign financial institutions.

You get the ABA routing number and SWIFT code from your bank. (Check the support pages at your bank's website: You're not the first person to need it.)

Once armed with this information, you can fill out the banking information form:

1. **On the Manage Your Contracts screen, under Bank Info, click Edit.**

   You land on the main Banking Information screen.

2. **To begin setting up your bank's information, click Add Address. Enter the mailing address for the bank where your account is held. Click Submit.**

   You should be able to get this information from your last bank statement.

3. **Enter your account information.**

   Select the currency used for the account, and then fill in the name fields for both the bank and the account holder. Choose the Account Type: You'll probably want to use a checking account so you can easily draw on the deposit.

   The remaining four fields are specific to your account. You probably already know your account number and branch. As noted above, you can also easily get the ABA number and SWIFT code.

4. **Double-check everything.**

   Really. Then do it again. If anything is wrong, you won't get paid, and it's a difficult manual process to get the information updated.

5. **Click Submit.**

   The last step is to certify that the information you entered is true and correct, and to authorize Apple to make payments into your account.

6. **Turn on the checkbox, check the information one last time, and then click Submit.**

Tax info

You're earning money and—no surprise—the government wants to know about it. The last step in setting up a Paid Applications contract is supplying your tax information. Since Apple is collecting payments for the sale of your goods, they're required by the IRS to report that income. They do this by electronically submitting a Form W-9 on your behalf:

1. On the main **Manage Your Contracts** screen, under **Tax Info**, click **Edit**.

2. **Enter your name as it's shown on your income tax return. If you have a business name that's different, enter that as well.**

3. **If you're a sole proprietor, select Individual from the drop-down menu. Otherwise, select the type of business entity that's appropriate.**

4. **For the Exempt Payee radio button, consult your accountant. You can download a PDF file of Form W-9 from the IRS website (*www.irs.gov*): It details the conditions for exemption.**

5. **Select your address from the drop-down menu. If necessary, click Add Address to specify a new address.**

6. **Finally, enter a TIN (Tax Identification Number).**

   If you're an individual, use your Social Security Number (SSN). For a business, it's the Employer Identification Number (EIN). Just type the numbers without any dashes or other characters.

7. **Turn on the checkbox to certify that the information is correct and then click Submit.**

Final approval

Your contract is ready for Apple to process. Back on the Manage Your Contracts screen, you see a green checkmark for Setup In Progress, but you still need to wait a few days for the information you submitted to be approved. Once everything has been sorted out, you see a new entry under "Your Contracts In Effect". Congratulations, you're now ready to earn money from iTunes!

Tip: Don't wait until the last minute to do this important electronic paperwork. If any problems arise with the information you've submitted, approval can take more than a few days and throw a wrench into your release schedule.

Below the contracts list are a few additional links for Canadian and Australian developers, who need to submit additional tax information. There's also a very important link that many developers overlook—Japanese tax treaty information. Not clicking this link can cost you a lot of money!

If you don't submit some additional forms to the Japanese tax authorities, your earnings will be subject to a 20 percent withholding rate. For every dollar you earn in Japan, 20 cents would be left behind. Submitting these forms is a Byzantine process that can take up to 90 days to complete, but in the end, it can be worth hundreds or even thousands of dollars each year. Remember to check out the link at the bottom of the Manage Your Contracts page once your contract goes into effect.

Stake Your Claim

Another iTunes Connect task you should do as soon as possible is to start adding your application's metadata. You need to provide the information that controls how your application is displayed in the App Store (both from within iTunes and from the app on the device).

It's possible to provide all this information in a single session, but it's easier and safer to work in small chunks. Don't be afraid to leave placeholder data the first time you add your application. You can make updates to the data as your release date approaches.

Note: The popularity of the App Store has caused iTunes Connect to be unresponsive at times. Working in batches reduces the risk of losing time and any data that's been entered.

Also, since metadata changes happen regularly, some developers track changes in a version control system. You can see an example of one such system if you look inside the iTunes folder of the Flashlight Pro project.

As you enter your information into iTunes, keep one important fact in mind: This data helps customers find and purchase your app. Make sure the information is clear, concise, and correct.

To get started entering your metadata, go to iTunes Connect's main screen Figure 8-1 and select Manage Your Applications. At upper left, click the Add New Application button.

Export Compliance

When you add a new application, the first question might seem odd: "Does your product contain encryption?"

Since iTunes will make your app available worldwide, Apple will export it outside of the United States. Special export requirements apply if your product has the ability to encrypt data. The link on the page provides additional information.

If your product relies on system facilities for managing encrypted data, such as the Keychain API to manage a user's private data or Secure Sockets Layer (SSL) for a network connection, you can select No. Apple has already authorized these parts of the iPhone OS for export.

For products that do contain encryption, you need to supply a copy of the commodity classification ruling (CCATS) as a part of the submission.

Overview

Adding a new application entails several screens of information. The first screen is Overview, which provides the application's basic information (Figure 8-2).

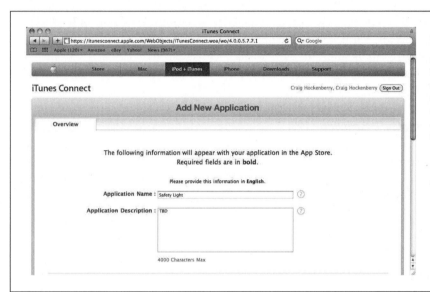

Figure 8-2:
The first thing you add to iTunes is an overview of your application. You have a lot of fields to fill out, but the following sections will help you through it.

Application Name

The first field in the overview is your application's name. This field must be unique: No products in the App Store have exactly the same name.

Some developers get around this by making subtle changes to the name. You'll find "Flashlight", "Flashlight.", "Flashlight .", and "Flashlight™" on the App Store. The problem with this approach is that it creates confusion for customers: If a friend recommends "Flashlight," which one is he talking about?

Note: This field is case insensitive, so "FLASHLIGHT" won't work. And you're left with a missed opportunity to be awesome.

It's also important to remember that the Application Name field doesn't need to match what's displayed on the iPhone's home screen. You could use a name like "Safety Light – World's Best Flashlight," but a name that long will get cut off in many of the App Store's screens.

The best approach is to keep your name short and simple. The name "Safety Light" best expresses this app's unique feature—the rescue beacon. Any promotions or reviews that mention the app by name make it easy for a potential customer to find it in iTunes.

Because your product name must be unique amongst hundreds of thousands of other names, it's a good idea to start adding your data into iTunes Connect as soon as possible. Some developers even create a placeholder application before they write a single line of code. Making sure that you get the name you want also prevents last-minute hiccups with your marketing efforts: Changing the name on promotional materials at the last minute can be costly!

Another thing to note about the application name is that you can't change it once it's in review. You can change it only when you update your application and upload a new version.

Application Description

Most developers set the initial application description to "TBD" (to be determined) or some other placeholder. What you put here depends on how you're going to market the product. In the early stages of the product's life in the iTunes database, you probably don't have that message clearly defined.

So you put it off until later. Just remember that you need to have a real description by the time you submit the final product for review. Your app won't get approved if your description is "TBD".

As far as the description goes, remember to keep it short and sweet. Only the first three lines of text show up on the initial screen. Many users in iTunes won't bother to click "More..." to see additional information. And a long description creates a lot of scrolling in the device's App Store application.

The text of the description can be any Unicode character. This means you can use symbols like bullets (•), stars (★), and other ornaments (♥ and ♣). Just remember that many of these characters look different on the desktop than they do on the device. Emoji glyphs are a good example: They don't appear at all in iTunes on the desktop.

Also avoid using prices in the description. In some cases, it can even lead to an app being rejected during review. The problem is that if you write, "PRICE REDUCED TO 99¢," it doesn't mean anything to someone in France, where the App Store is displaying prices in Euros.

Tip: Do yourself and your potential customers a favor, and check the spelling before you submit your description.

If you're updating an application description because you're releasing a new version of the application, remember that the information will be updated in iTunes on that same day, not on the day that the new version is approved. If you're going to list additional features, do the update after they're available in iTunes.

Device Capabilities

After providing your application description, you see a question: "Do you want to limit your app to only run on devices with specific capabilities?" You need to answer Yes or No, but the answer really doesn't matter.

If you click the Yes radio button, you see a link to some online documentation that describes how the *UIRequiredDeviceCapabilities* key in the Info.plist file works. The buttons don't affect your application's information in any way; it's just Apple's way to get you to Read The Fine Print?

When the *UIRequiredDeviceCapabilities* key isn't listed in your app's Info.plist, all customers on iTunes will be able to download it. That's the best goal to have, because the greatest number of people will be able to buy and enjoy your app. But sometimes your creation will rely on specific hardware features.

For those cases, you need to specify an array of strings that lists the items required by your application. For example, you could specify the following in your Info.plist:

```
<key>UIRequiredDeviceCapabilities</key>
<array>
    <string>still-camera</string>
    <string>magnetometer</string>
</array>
```

The *still-camera* string says that your application won't function correctly unless it can take pictures. iPod touch owners won't be able to launch the application, because that device doesn't have the camera. Similarly, the *magnetometer* string indicates that your app needs the ability to determine a heading. A user with an older iPhone won't be able to use the app, because the compass only became available on the 3GS model.

Tip: To learn more about device capabilities and to see a full list of capability strings, use the documentation viewer to search for *UIRequiredDeviceCapabilities*.

When you upload your application binary to iTunes, the contents of the Info.plist file will be examined. This final step in the App Store submission process stores the hardware requirements along with your other application data. This information will prevent customers from installing the application on a device that does not meet the requirements.

You may be tempted to just leave out the device capabilities information and let all customers download and install your application. You're assured of having the largest number of potential buyers, but it also ensures you get the most customer support problems if there's an incompatibility.

Before making any decisions about your app's hardware requirements, make sure you've tested the application on a range of devices—even if it means that you need to get on eBay to buy a first-generation iPod touch.

Categories

The Primary Category menu determines where your application appears in iTunes. Your app will appear as a new release and in the Top Downloads lists for the selected category. In some cases, the category is also displayed below the app icon and name.

The Secondary Category is optional and only comes into play with the iTunes Power Search feature. Your application won't be listed in both categories. Set it to something relevant, but don't expect it to double the number of customers who see your product.

Copyright and Version

The next fields are the copyright and version that will be displayed on the product page. For the copyright, enter the year and your company name: "2010 Squidger Industries", for example. If you don't add a © symbol, one will be added for you at the beginning of the text.

The version should be just numbers and periods: values like "1.0" or "1.0.1", not "1.0 final" or "Version 1.0".

SKU Number

The SKU (Stock Keeping Unit) number can be any sequence of numbers and letters you'd like. The only requirement is that this value is unique to your developer account: You can't have two products that use "OMGWTFBBQ001" as the SKU.

Be careful about what you choose for this field: Once a value has been submitted, you can't change it under any circumstances.

You'll find that this SKU number appears as a vendor identifier in the end-of-month financial reports. You may want to use the value to sort or filter data, so take that into consideration when making your choice.

Keywords

The keywords, along with the application and company name, are how iTunes finds app matches when users search. Without keywords, a search for "flashlight" or "SOS" wouldn't return any results with the Safety Light application.

When entering keywords, separate each word with a comma. The number of characters in the field is limited to 100 total, so choose your keywords wisely. A word processor can be helpful for checking the length of your keywords: Bean is a good choice because it displays the character count at the bottom of the window as you type. You can download it for free: *http://www.bean-osx.com/*.

As with the application name, you can edit your keywords up until the final binary is submitted. Once that file is uploaded, the keywords are fixed in place until your next version. Think carefully about the words you put in your application description. Try to put yourself in the customers' shoes; think about how they might search for your application.

Avoid some keywords: Any offensive words or trademarked names will cause your application to be rejected. Likewise, if you use the names of other popular applications as a way to fool users into finding your product, you'll run afoul of the review process. If you use common sense and don't try to game the system, you'll be fine.

URLs and Email

The next group of input fields is for your URLs and support email address. The Application URL field isn't required, but it's highly recommended, since it's the first one that appears in iTunes (under the application's description). The text label for the URL is generated automatically using your iPhone Developer account information. If you have a company account, you see something like "Squidger Industries Web Site", while individual accounts, for example, would display "Al Czervik Web Site". If you have a website that shows all the products you sell in the App Store, it's a good candidate for this field.

A Support URL, on the other hand, is required and should point to a page where customers can get more information about your product. Also have an email address or some type of contact form on that web page—it'll be the only way customers can contact you.

Tip: The Support URL also appears in the App Store application on the device. Unfortunately, it's not a clickable link, so it's harder (and less likely) for a user to visit your product web page.

But what about the next field, Support Email Address? Can't customers send you email using the value you enter there? Unfortunately, Apple only uses this email address in the event that *they* need to contact you about the application.

Demo Account

Another optional field is "Demo Account – Full Access". The information you supply here is to help Apple's review staff evaluate your application. If your app communicates with a web service that requires a login, such as an online scoreboard or a social network, you need to provide access information for the reviewers.

EULA

Some apps need a custom End User License Agreement (EULA). Links at the bottom of the Overview screen let you specify any special terms required by your legal department.

The license agreement that you submit can't have any markup—just text and line breaks. If you supply a custom EULA, a link labeled "Application License Agreement" appears next to the support link on the product page. In the App Store application a similar link is at the bottom of the scrolling list.

If you don't specify a EULA, you'll be covered by Apple's standard license agreement. (There's a link to view this document at the bottom of the Overview screen.)

Before moving on

You need to supply data for all the required fields before you click the Continue button at the bottom of the screen. Again, if you don't have all the necessary data, you can supply "TBD" or some other placeholder. All of the values, except the version and SKU numbers, remain editable until you submit the final binary.

If any validation errors occur, they appear at the top of the screen in red. The most likely error is that the application name is already in use by another developer. If you see this error, try submitting another one. This process is as much a marketing exercise as a technical one, so make sure to get input from other people in your organization.

Ratings

After you've completed the overview, the next step is to rate your application. If your application contains any obscene, pornographic, offensive, defamatory, or other content that Apple doesn't approve of, it's not going to be listed in the App Store (Figure 8-3).

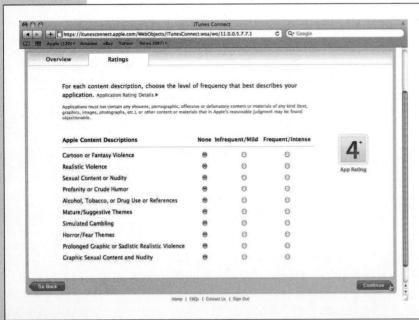

Figure 8-3:
The iPhone and iPod touch both have parental controls. By rating your app, you're letting parents restrict content inappropriate for their children.

Even if your content isn't objectionable, it still needs to be rated. That's because the iPhone OS has parental controls. Parents can configure, for example, a child's iPod touch to install only content with the appropriate rating. To see how these controls work, open the Settings app on your iPhone and navigate to General→Restrictions. Once Restrictions are turned on, a parent can specify "Don't Allow Apps", "4 to 17 years or older", or "Allow All Apps in the Allowed Content" section. Entering a passcode is required for the user to change these settings.

Setting the rating is simple: Just select the radio buttons according to the different kinds of content in your app. Each type of content will either be "None", "Infrequent/Mild", or "Frequent/Intense", and the overall rating will be updated as you change the selections.

If you see "No Rating", Apple won't allow the application in the store. If you're skirting on the edges of what's acceptable, it's a good idea to experiment with these radio buttons while you're designing the app. As with the application name and keywords, you can't change ratings once your application binary is submitted. You can update them with each new version you submit.

Once you're satisfied with your rating, click Continue to advance to the next screen.

Upload—Show It Off

Now's the time to really show off your app! The Upload tab (Figure 8-4) is where you submit your application icon and screenshots that customers will use to check out your app before they make a purchase.

Figure 8-4:
The Upload tab in iTunes Connect lets you submit your application icon and other graphics. You can click the preview links under each item to make sure the file transferred correctly. Also make sure that the "Upload application binary later" checkbox is turned on.

Application binary

Skip uploading the first file, Application, by turning on the "Upload application binary later" checkbox, for these reasons:

- **You want to get your metadata into iTunes as soon as possible.** That means you probably don't even have an application binary to submit: It's not uncommon to start setting things up even before you start your beta test!

- **iTunes Connect has a history of being unreliable.** A combination of many developers and high loads on an evolving system results in more downtime than anyone likes. It can take awhile to upload your application binary, and if any kind of problem arises during that time, you lose the other metadata you've entered.

The only items that are required on this page are the "Large 512 × 512 Icon" and the Primary Screenshot. And even if you don't have final versions of these files, you can upload a placeholder graphic that can be replaced as your release date approaches.

Application icon

The application icon graphic is much larger than the 57 × 57-pixel icon that you use in the code. If your designer produced your icon with illustration software that produces vector graphics, it will be a simple matter to scale up that image. If not, she'll have some extra work to do. You don't want to scale up a non-vector graphic because it'll get pixilated and blurry; that's a terrible first impression to make with your customers.

Of course, the advantage of using this separate graphic is that it doesn't have to be exactly the same as the one that appears on the iPhone home screen. Many developers put "ON SALE" and other sales messages on this file during their promotions. Just keep in mind that this 512 × 512 graphic will be scaled down significantly. Make sure any text or details look OK after being reduced in size.

This graphic is used in iTunes only on the desktop. The App Store on the device uses the same 57 × 57-pixel icon from the application. All promotional badges that you use won't be seen by a majority of potential customers.

Tip: You may even want to include this graphic as *iTunesArtwork* in your beta test build. As explained in the note on page 225, the Flashlight Pro project shows how to do so.

To upload your icon, click the Choose File button, and navigate to your TIFF or JPEG file. This file should not contain any layers or transparency.

After you upload the graphic successfully, you see a checkmark icon. You also see the file name displayed along with a link that opens the graphic in a new window. Click the link to verify that the image uploaded correctly.

Screenshots

Uploading the Primary Screenshot is identical to the process used for the application icon. The only difference is the size of the files that can be uploaded.

You can submit screenshots either with or without a status bar at the top: 320 × 480 or 300 × 480 pixels in TIFF or JPEG format. The files can be in portrait or landscape orientation: iTunes and the App Store application position the files correctly when they're displayed on your product page. This image is always displayed first on your product page.

Uploading Additional Screenshots is optional, but you're crazy if you don't use them. Most customers use these pictures to evaluate the suitability of your app. The more visual information you can provide, the better.

The upload process for these additional screenshots is a little confusing. You click Choose File to select each of the images, and then click Upload File to submit them all at once. The tricky part is getting the screenshots in the correct order. You're trying to tell a story about your application, and if the storyboard images are in the wrong order, the plot is confusing.

Another common mistake developers make is that they present the screenshots in the order that they appear in the application. People don't want to see a game setup or login screen—they want to see some action and the cool stuff your application does. If you have help screens, these can also be good candidates, because they give an idea of how your application works in a graphically rich way.

Once you figure out the screenshot order you want, choose the files in the *opposite* order in which you want them to appear in iTunes. Say you have four additional screenshots:

1. **Click Choose File and select your fourth (last) screenshot in the file dialog box; click Choose.**

2. **Click Choose File again and repeat the process with your third screenshot. Then do the same thing for the second and first additional screenshots.**

   The files should then be listed in the correct order above the Upload File button.

3. **If so, click the Upload File button**

4. **iTunes Connect transfers the files to iTunes. You can click Clear All and start again.**

Tip: The easiest way to keep your sanity here is by naming the screenshots with a number. Your primary screenshot is *0.jpg,* the first additional screenshot is *1.jpg,* and so on. This makes it easy to see that *1.jpg* is followed by *2.jpg* as you choose the files.

Click Continue and you're done uploading files for your application. Remember that you can come back to this screen at any time and update the images, even after your product is released in the App Store.

Pricing—Pick Your Price

Only two fields are on the Pricing tab (Figure 8-5), but are they important and hard to get right!

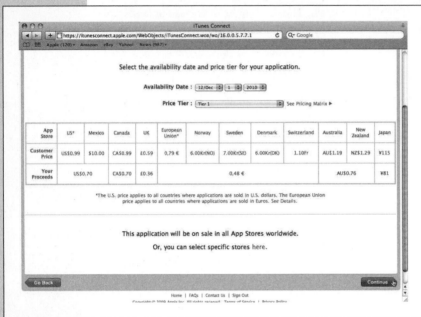

Figure 8-5:
The Pricing tab in iTunes Connect only asks for two pieces of information, and they're both very important. The Availability Date controls when the product is released and should initially be set to a date far in the future. The Price Tier selection controls how much you earn from each sale. The tab displays a summary in each currency.

Availability Date

iTunes' default behavior is to make your new release available immediately. To control the date your app is released, select an availability date that's well into the future. Give yourself enough time to choose a specific date and to coordinate your marketing efforts around it. Consider a time to send out review copies, get press releases distributed, and to deploy your promotional site. If the switch gets thrown at some random date in the middle of the night, you'll end up with a dud instead of a launch. By setting the date in the future, you'll have a chance to get everything organized and to update the setting to your actual launch date.

Another important factor when choosing a launch date: It won't happen on the day you think it does. It's easy to forget that iTunes is a worldwide enterprise. You may be living in California, but plenty of your customers are in New Zealand and are a heck of a lot closer to the International Date Line than you are.

When you specify Availability Date as "10/Oct - 22 – 2009", you may be surprised when people start hitting your website and sending you support email at 4 a.m. on the day before the launch. iTunes made your application available in New Zealand on the 22nd as you requested, but you didn't know that New Zealand is 20 hours ahead of your local time. Whoops!

To figure out when the launch will really happen, use an online time zone converter: *http://www.timeanddate.com/worldclock/converter.html.* Select midnight in Auckland, New Zealand, as the time and place to convert from, and choose your location as the place to convert to. If you're in the United States, it's going to be sometime in the early morning. Plan your sleep schedule accordingly.

Price Tier

After selecting the availability date, pick a price for your app. Again, since iTunes works worldwide, you don't set the price in dollars and cents. Instead, Apple uses *Price Tiers*—groups of equivalent prices.

The first option is Free, and selecting it means that you won't make any money off the download of the application. (You can still make money with in-app purchases and advertising, however.)

If you pick Tier 1, your application will go for 99¢ in the United States, €0.79 throughout Europe, and the equivalent prices in other foreign currencies. Each tier level is approximately equal to one dollar until you get to the higher ones. (Tier 85 means $999.99 in the United States!) After you select a tier, the amount in various currencies appears along with your proceeds from each sale.

Of course the hardest part is figuring out how much you should charge for your application. The next section will cover some of the market factors that may influence your decision. For now, just pick a tier that you think is close, and click Continue to advance.

Localization

You've worked hard to write a great application description, so pick relevant keywords, and showcase the best images from your app. And don't forget languages other than English. Otherwise, up to 40 percent of your potential customers may never discover your app. When someone in Spain searches for *literna,* you want to make sure he finds your flashlight!

Again, to be successful in iTunes, you need to think globally. A large portion of your customer base doesn't speak the same language you do. You only have their attention for a short time: Make your marketing message easy to read.

The Localization tab lets you supply translations for the following items:

- Application Name and Description
- Keywords
- URLs and Email
- Primary and Additional Screenshots

iTunes Connect supports many different localizations, but you get the most bang for your buck by translating your marketing materials into Japanese, French, German, Spanish, and Italian.

To enter a translation, select one of the languages from the pop-up menu (Figure 8-6), and enter any values that need to change. If you don't specify otherwise, the screenshot images from the English version display on your product page. Once you've entered the localized values, click Add Another Language to go back to the language selection menu, or click Continue to finish entering the metadata.

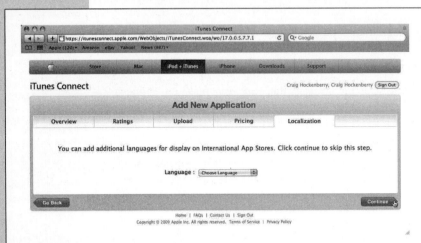

Figure 8-6:
Use the Localization tab to provide your application information in different languages. The pop-up menu selects the language and presents you with more text input fields.

Review

The last step is to check your information. The tab's default view shows how the data will appear in the United States. You can, and should, change the pop-up menu for Review App Store to other countries to see how your marketing information will look for customers using that store.

To make any changes, click the Go Back button at the bottom of the page, or click one of the tabs at the top of the page. If everything looks hunky-dory, click Submit Application, and your product will be added to the iTunes database. Congratulations!

Tweak It

As you've seen in this section, setting up your metadata is an iterative process. Once you've entered the basic information for the application, you can click its icon in the Manage Your Applications section of iTunes Connect. You see a screen with an Edit Information button that lets you update the metadata.

Figure 8-7:
The final step in submitting your information to iTunes is to review it for accuracy. You can use the Review App Store pop-up menu to preview the information on App Stores throughout the world (A). Check the spelling in your application description, and make sure that your categories, keywords, and URLs are all correct (B). You can click the links for the images to view them in a new window (C). Last but not least, your Availability Date and Price Tier specify when the product will go on sale and how much it will cost (D).

Note: As your app goes through its normal lifecycle, the metadata changes often. In fact, you'll probably want to maintain a document in your source code tree that all members of your development team can update and revert as necessary.

The Flashlight Pro project keeps all the iTunes information in the root of the project folder. The folder contains an AppInfo.rtf file with metadata along with the application icon and screenshot images.

The Market

Like the platform itself, selling products for the iPhone is a whole new world. Chances are you've never sold software to a mass market: That's about to change with iTunes!

As discussed on page 258, setting your application's price is one of the more difficult aspects of preparing your release. You're the only one who knows how much work went into the product and how many licenses you need to sell in order to recoup your investment.

To help you make this decision, you need to look at the market you're about to sell into. In Chapter 4 (page 113), you took the first step in marketing: thinking about how to position your application before you even start coding. Now it's time to start thinking about how you're going to price and then promote that final product.

Metrics

To begin to understand this new market, take a look at some of the numbers that define it. Keep in mind that this new market is also evolving quickly—the links included below should help you keep up with the latest changes.

Surprise! Apple is secretive

Apple doesn't provide much information about the iPhone OS market. The only substantive metric is the number of iPhones sold in each quarterly financial report. Using these numbers, you can see in Figure 8-8 that growth of the market is pretty impressive.

It's interesting that these numbers don't include a significant portion of your potential customer base: Apple does not break out the numbers for the iPod touch (instead, it reports the total number of iPods sold). The only information from the company regarding the breakdown was during their quarterly earnings call in July 2009. They hinted that 45 million devices were in circulation, suggesting that the iPod touch makes up a little over 40 percent of the total market.

Third-party research

You can get additional market information from third parties. Companies like AdMob gather data from ads and surveys they deliver to the devices. Other companies, like Pinch Media, let developers add analytics libraries that collect anonymous statistics. Even established research companies like the Nielsen Company are surveying users in order to establish market patterns.

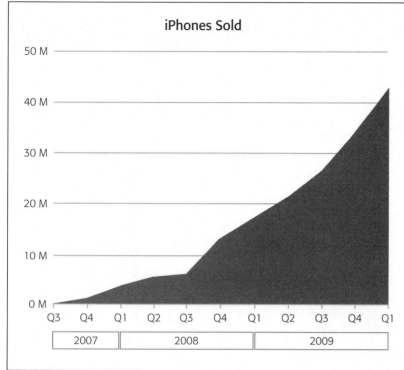

Figure 8-8:
This graph shows the cumulative number of iPhones sold since the middle of 2007. The sales growth appears to be exponential, causing some analysts to predict over 70 million devices by the end of 2009. Note that these sales numbers from Apple don't include the iPod touch, and that their fiscal year ends in September.

These sources have uncovered some interesting results:

- Ninety percent of users download apps directly from their phone (avoiding iTunes on the desktop).

- iPhone users download about 10 new apps each month, about 25 percent of which are paid. On the iPod touch, users pay for about 10 percent of 18 apps downloaded each month.

- The average user spends about $9 on five paid applications each month. That averages out to $1.80 per app.

- Seventy-five percent of all users have downloaded an app (free or paid).

- Only 20 percent of users continue to use a free application the day after it's downloaded. After 1 month, only 5 percent of users still use the application.

Here's where you can check out these sources for yourself. You can find clickable links on this book's Missing CD page (*www.missingmanuals.com/cds*):

- **AdMob.** *http://metrics.AdMob.com/2009/08/july-2009-metrics-report/*

- **Nielsen.** *http://blog.nielsen.com/nielsenwire/online_mobile/iphone-users-watch-more-video-and-are-older-than-you-think/*

- **Pinch Media.** *http://www.pinchmedia.com/appstore-secrets/*

User demographics

In June 2009, AdMob and comScore conducted a survey to establish the demographics for iPhone and iPod touch users. Not surprisingly, iPhone users are generally older, while those on the iPod touch tend to be younger (Figure 8-9).

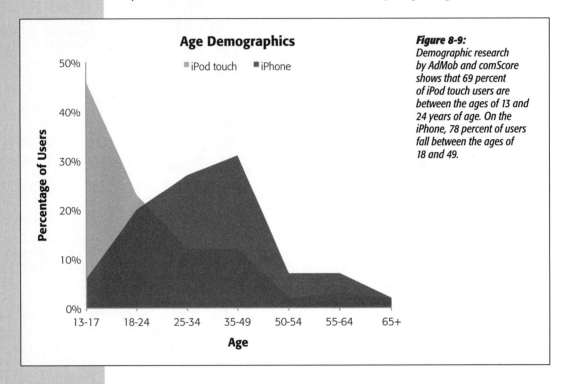

Figure 8-9:
Demographic research by AdMob and comScore shows that 69 percent of iPod touch users are between the ages of 13 and 24 years of age. On the iPhone, 78 percent of users fall between the ages of 18 and 49.

Along with the older age demographic, iPhone users earn more. Only 66 percent of iPod touch owners made over $25,000 each year versus 78 percent of users with iPhones. This survey also reported that over 70 percent of users are male. A prior comScore survey in October 2008 showed that 43 percent of iPhone owners earn in excess of $100,000. More interesting was the strong growth of iPhone ownership below the median household income: 48 percent in the $25,000–50,000 bracket and 46 percent in the $50,000–75,000 bracket.

This strong growth during a global economic recession suggests that lower income users don't view the iPhone as a luxury purchase. Instead, it's viewed as a single communications and entertainment device that is cheaper to own and operate than multiple devices and services.

Sources:

- **AdMob.** *http://blog.AdMob.com/2009/06/16/new-research-on-the-demographics-and-behavioral-characteristics-of-iphone-and-ipod-touch-users-from-AdMob-and-comscore/*

- **comScore.** *http://www.comscore.com/Press_Events/Press_Releases/2008/10/Lower_Income_Mobile_Consumers_use_Iphone/*

Don't forget, the average iPhone customer is also paying thousands of dollars over the life of his contract. The research highlighted in this section shows that iPhone and iPod touch owners have good incomes with plenty of disposable income. The size of the customer base is also growing rapidly across many demographics.

That's great news for you as an iPhone developer! Plenty of customers are waiting to acquire your product via iTunes. There's just one problem: a rapidly growing market filled with lots of willing customers leads to…

Competition

You're not the only developer who's attracted to this new market. In a recent response to an FCC inquiry, Apple reported that approximately 6,800 iPhone applications are approved each week. Granted, many of those applications are updates to existing titles, but the App Store still gets over 10,000 new applications each month.

Sources:

- **Daring Fireball.** *http://daringfireball.net/2009/08/apples_fcc_response*
- **148 Apps.** *http://148apps.biz/app-store-metrics/?mpage=appcount*

Running up the charts

If you're considering making a run at the charts, you're going to find it very hard to do. With the increased number of titles available comes an increase in the number of sales needed to enter the Top 100.

When the App Store debuted with approximately 1,000 applications, you needed to sell about 100 copies of your app per day to get into the chart. Now, with over 100,000 different apps, you need to sell over 1,200 copies per day. (For the full story, check out *www.taptaptap.com/blog/the-app-store-pricing-game/*.)

You'll also find that the requirements to get a chart-topping application are much higher than they used to be. Long gone are the days where you could make a farting or flashlight app in a few hours and then rake in tens of thousands of dollars from it being a hit. Developers have learned that higher quality and customer value are a competitive advantage, and they spend months developing titles that they hope are hits.

Similarly, getting featured in the "New & Noteworthy" or "Staff Favorites" sections of iTunes becomes less likely as the number of titles in the App Store increases. If it happens, great, but campaigning to get featured shouldn't be the main focus of your marketing efforts.

Press

Another facet to the market saturation is that it's becoming increasingly difficult to get press for your application. Online media is having just as hard a time keeping up with demand as the rest of the iPhone ecosystem. Don't wait until the last minute to get the press interested in your work. Your marketing efforts should begin well ahead of your release date. Remember that anyone you contact is up to her eyeballs in press releases. Look for angles that will help the writer come up with fresh content. Maybe your app fits into an upcoming holiday, or it does something no other app has done before. (It could even be a flashlight that announces a new book about iPhone development.)

As you look for sites to feature your application, don't be afraid to target a niche. It's unlikely that IGN will be interested in Tiddlywinks Pro. Spend a few minutes with Google and search for *Tiddlywinks blog*, and you'll find plenty of places that would be thrilled to learn about your cool new iPhone game.

Finally, be warned that some "review" sites are charging developers for increased attention. Don't go anywhere near these sites—word is quickly spreading that the value of the recommendation isn't worth the paper it's printed on. That's saying something, considering it's a web page!

Two Developers, One Store

As a developer, you can take two basic approaches to selling your app. You can enter the new and fast-paced world of hit-based marketing, or you can take a more traditional approach of slowly building a loyal customer base.

The germ of this idea came from Marco Arment (*www.marco.org/208454730*), developer of the Instapaper app, and John Casasanta (*www.taptaptap.com/blog/convert-first-month-sales/*), who led the team that developed Convert and other iPhone hits. Each has been successful selling with one of these two approaches.

To see which approach is best for you, check out Table 8-1.

Table 8-1. *Two ways of positioning your application in the App Store.*

	Hit-based	Traditional
App Types	Simple and shallow.	Complex and deep.
Typical Categories	Games, Utilities.	Productivity, Photography, Medical.
Customers	Scan "top" charts looking for titles. Rarely look for apps outside of iTunes or App Store application.	Research on Web for available apps before making an informed purchase. Read positive review. Clicks advertisement.
Characteristics	Novelties with broad mass-market appeal. Apps have a limited shelf life, often just months.	Narrower appeal for individual tastes and needs. App becomes a part of customer's daily life.
Price Points	Lower prices to encourage impulse buying.	Higher prices from customers who appreciate depth and quality.
Marketing	Huge PR blitz at launch to generate short-term buzz about new product.	Relies on good press, positive reviews, and word-of-mouth recommendations for long-term product interest.
Launch Pricing	Low introductory prices to increase sales volume and move up the charts.	No introductory pricing, which leads to bad evaluations from customers who aren't invested in product.
Income Potential	High, but comes at risk of not recovering initial development costs.	Lower, but much better chance of income for sustained development.

Keep in mind that these aren't hard-and-fast rules, just generalizations that will help you figure out what works best for your products.

Look at the music business

When using the hit-based approach for selling your apps, it's just like being in the music business: You live and die by the charts. If you look at the world of pop music, you see that most songs don't spend much time at the top. After a month or two, the public moves onto something else and sales languish. While a song is on the charts, a lot of money is spent on promotion to keep the track listed.

And even if you make it onto the charts once, many musicians find that they're one-hit wonders. It's hard enough to get the first single on the charts: Continued success is even more elusive.

At the same time, a lot of musicians make a great living without ever being on the Billboard Hot 100. Some of them do session work, while others release albums to a loyal group of fans. The App Store business is no different.

Choose your customers

When you're selling to a mass market, you deal with two types of consumer behavior. Some customers want the cheapest solution that gets the job done and don't care about elegance. If they see a product for $5 and another for $10, they always choose the lower priced one.

Other customers want to pay more and get the best there is. When presented with the same choice between a $5 and $10 product, they assume the more expensive one is better.

This consumer behavior tends to affect your enterprise: If you sell to tightwads who have no brand loyalty, your business will reflect that. Alternatively, if you work at finding great customers who want long-term success for your products, your business will grow with that investment.

High roller

If entering the hit-based market feels a lot like gambling, that's because it is. Spending a lot of money on a marketing blitz increases your chance of success, but it doesn't ensure that you'll have a hit. You could make millions in your first year, but odds are that you won't.

The alternative of using a more traditional marketing approach lets you establish a steady income with sustainable pricing. You may not make millions in your first year, but you build a long-term business with loyal customers.

Pricing Strategies

All of this talk about the market, but what about the topic of prices? It's the final piece of a complex puzzle.

A cup of coffee

The price of your product depends a lot on what it does, not how hard it was for you to make it. A lot of frustrated developers trot out the line that "my app is worth more than a cup of coffee!"

From your point of view, that's an entirely valid statement. You know how much work went into bringing it to life. Unfortunately, your potential customers don't care about that. In their eyes, a good cup of coffee is worth $5. Your app isn't.

The trick is to increase your product's perceived value.

What kind of product is it?

The price you set implies a lot about the kind of product you're selling. The following price ranges are typical in the App Store:

- **$0.99.** The lowest price is where you find novelty applications that provide entertainment value or simple utility. It's also used for promotions of apps that normally carry a higher price.

- **$1.99 to $2.99.** These prices are typically the upper bound for an app that you want someone to buy on impulse.

- **$3.99 to $4.99.** This price range works for premium apps that have demonstrable high quality. This is also the level where customers start to experience price sensitivity.

- **$9.99 and above.** Specialty applications can attract customers who are willing to pay much more than average. If you're targeting specialized needs in a vertical market, selling the app as an enhancement to an existing product, or providing access to an expensive dataset, typical App Store pricing won't limit you.

Consumers aren't biased toward one price point over another. Data collected by Pinch Media shows the average number of paid downloads is consistent for all prices below $4.99. Their research also shows that free applications are used less frequently than paid ones, which suggests that users form an attachment to their purchases.

One thing to keep in mind as you set a price: The higher you go, the more likely you are to garner positive ratings and reviews in iTunes. As people spend more money on your product, they're also more likely to think that the purchase was a good one. This form of rationalization helps the customer avoid *cognitive dissonance*.

In turn, positive evaluations will influence other customers. Given an uncertain situation, the psychological phenomenon of *social proof* takes hold: People tend to look to others for appropriate behavior. In the App Store, ratings and reviews provide those clues. It's much easier to make a sale if previous buyers have given it their approval.

Positive reviews also play an important part in the early days of a product's release: They help generate momentum. Don't be too proud to ask friends, relatives, and followers on Twitter to give you a review. Most will be happy to help you get established!

Sources:

- **Cognitive dissonance.** *http://en.wikipedia.org/wiki/Cognitive_dissonance*

- **Social proof.** *http://en.wikipedia.org/wiki/Social_proof*

- **Mobile Orchard.** *www.mobileorchard.com/app-store-heresies-higher-price -better-ratings-dont-discount-your-app-at-launch/*

- **Pinch Media.** *www.pinchmedia.com/blog/paid-applications-on-the-app-store-from-360idev/*

Try Before You Buy

From a customer's point of view, the best way to know if you like a product is to actually use it. If you're looking for a phone, you can walk into an Apple Store, pick up an iPhone, and play around with it. In a matter of minutes, you know if that product suits you.

The same is true with applications in iTunes. Many developers have found success by providing a free version of their application. A customer can then try the application with no obligation and upgrade to the paid version if they find the application useful.

Incentives

For this technique to work, you need to give the user an incentive to upgrade. Positive motivation works best: You offer the user more features, additional game levels, and other content that enhances their experience. Users will pay for the added value.

You can also use negative stimulus to convert the free user into a paid one. In this case, "negative stimulus" is another term for "advertising." Ads are annoying and take up precious screen real estate, so some people are willing to pay to get rid of them.

Using advertising as an incentive has a side benefit: Ad revenue supplements your income while people evaluate your app. In some cases, the money earned from advertising alone is enough to pay for development efforts.

Advertising isn't appropriate for all types of applications. It works best with apps that get regular use: advertisers like repeated exposure. You typically get paid according to how many ads you show, so repeated use increases your earnings.

Conversion rates

So how many of these free downloads will result in a paid download? It varies greatly by application. Some developers have reported a conversion rate as low as 1 percent, while others have seen it as high as 15 percent. Pinch Media has reported an average conversion rate of 7.4 percent across a large sample of applications.

On average, for every 14 people who download your free version, 1 person will pay for it. That doesn't sound great until you remember that iTunes has millions of accounts!

Free to paid

Once users decide to upgrade to the paid version, how do they do it?

Traditionally, developers have submitted two versions of their applications to iTunes. One is a "Free" or "Lite" version, and the other is a "Premium" or "Pro" version. Be careful how you brand both of these versions; make it clear to the customer that one version is a subset of the other.

Apple also requires that free demo versions be fully functional. You can't have any features that get disabled over time. These free applications must also be useful in their own right. The customer should not feel forced to upgrade.

Several problems come with maintaining two versions of the application. First you're doubling the amount of work it takes to build, test, and submit an application. You may also need to deal with data migration issues. Since apps can't access each other's data, you need to find a way to get any documents, scores, or other information from the free version into the paid version.

A recent policy change by Apple provides a much better solution for you and your customers: A free application can use in-app purchases to unlock paid features. With this technique, you provide a unique product ID to iTunes by using the StoreKit framework. Once the user approves the transaction, get information that lets you unlock features. You don't need to manage two applications, there are no data migration issues, and it's a much simpler experience for your customer.

Upload

Once you conclude your beta test, you have to take one more pass at adjusting your Xcode settings. You signed the distribution builds for your beta test by using an Ad Hoc provisioning profile (page 215). The build you submit to iTunes should be signed with an App Store profile. This section shows you how to create the new App Store profile, create and upload the new build, and how to test it. Last but not least, you'll start to promote your new app and get ready for buyers.

The Final Profile

To create the new profile, fire up your old friend the Provisioning Portal website (page 136). Then follow these steps:

1. After logging into the Provisioning Portal, select Provisioning in the left column.

2. On the Provisioning page, click the Distribution tab. On the Distribution tab, click New Profile to create a new distribution profile.

3. For Distribution Method, leave App Store selected.

4. Next, supply a Profile Name. If you named your beta profile Squidger Beta Distribution, then a good name would be Squidger Final Distribution.

5. Select an App ID for the profile. Use the same one you used for your beta test: "Squidger Apps", for example.

6. Since you're going to distribute this version on the App Store, you don't need to specify devices. Just click Submit, and your new profile will be created.

Click the Download link to retrieve the profile, and then drag it onto the Xcode icon in the Dock to install it. You should see "Squidger Final Distribution" in the Organizer window.

The Final Target

You'll want to create a new target for this build, much as you did with the beta test build (page 127). This target will be just like the one you did for your beta test, but will have different settings in the Code Signing build settings.

For the beta test, you added a *dist.plist* file for the Code Signing Entitlements. For the final release, you do *not* need this file. The other setting, Code Signing Identity, should be set to Any iPhone OS Device. In addition, you do *not* want to use the automatic profile selector, because it may select your beta profile by accident. Instead, scroll down in the list, and select the iPhone Distribution that's listed under the profile name (Squidger Final Distribution, for example).

To see these settings in action, look at the "Safety Light (Final)" build settings. Double-click the target, and you see no entitlements and that the code-signing identity is "iPhone Distribution: Craig Hockenberry" followed by a long and globally unique ID (GUID).

You see this GUID because you don't have the distribution profile or code-signing certificate for the "Craig Hockenberry" developer account. If you try to build the project, you see a code-sign error stating that there isn't a "valid certificate/private key pair in the default keychain". If you want to submit this application to the App Store, you have to use your own unique certificate and provisioning.

Tip: If you see a GUID, it generally means the provisioning profile can't be located by Xcode. Download a new copy from the Provisioning Portal.

The Final Build

Now that you have your target configured for distribution on the App Store, you're ready to do the build that all your customers on iTunes will use. Before you do so, make sure to update the "Bundle version" number (page 221) in your Info.plist to use just numbers and periods—no letters. If you leave in a beta version number, like "1.0b3", it'll be rejected on upload.

After the build, you need to make a backup copy of both your application and the dSYM file. Both files are *required* to investigate crashes encountered by your customers. (You learn how to debug these crashes on page 298.)

You also need to archive the app you just built in a Zip file before uploading it to iTunes. (Simply Control-click the file, and choose Compress from the shortcut menu.)

Tip: The Flashlight Pro application has an aggregate target named Final with a Package Release build phase that creates the Zip file automatically. This build phase also creates a backup of the app and dSYM files so they can be checked into your version control system.

The Final Upload

Once you have the final Zip file ready to go, it's a simple matter of uploading it with iTunes Connect:

1. **Log into iTunes Connect (page 233) and click Manage Your Applications. In the list that appears, click your application icon.**

   Your application's Status should be Waiting For Upload.

2. **In the application information screen, click Upload Binary to start the upload process.**

3. **On the next screen, click Choose File, and then navigate to the Zip file that contains your final application. When you're done, click Choose.**

4. **A new button appears on the screen: Upload File. Click it, and the file uploads to Apple's server.**

5. **When the upload completes, you see a checkmark icon; click Save Changes to complete the submission process.**

   At this point, the server checks your uploaded file (Figure 8-10). If it finds any problems, you see a red banner at the top of the page that explains the error. A common source of problems is the distribution provisioning: Go back and check your build settings if you encounter this type of problem.

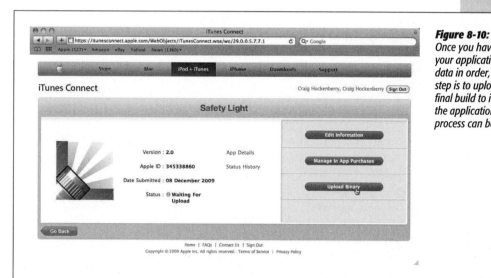

Figure 8-10:
Once you have all your application metadata in order, the last step is to upload your final build to iTunes so the application review process can begin.

Note: You may also see an error like "We're temporarily unable to save your changes. Please try again later." Sometimes the system gets overloaded, and you have to try several times for it to work. For the same reason, update your application metadata separately from uploading the file.

If everything goes fine, your application's status changes to Waiting For Review. It's just a matter of time now. Congratulations!

The Final Test

There is, however, a slight problem: You've never run the final build that you uploaded to iTunes. Nor can you, because it's signed with an App Store distribution profile that has no devices specified.

In just a short time, hundreds or even thousands of customers will be buying and using this build, so it sure would be nice to know that everything is OK. Fortunately, you can get that reassurance by using a command-line tool called *codesign*. This tool uses the same methods as Xcode uses to sign code.

After changing your directory to where you did the build, you can view the current values for code signing in the app with:

```
$ cd build/Final--Release-iphoneos
$ codesign -d -vv "Safety Light.app/Safety Light"
Executable=/Users/studmuffin/Projects/Flashlight Pro/build/Final--Release-
    iphoneos/Safety Light.app/Safety Light
Identifier=com.iconfactory.FlashlightPro
Format=bundle with Mach-O thin (armv6)
CodeDirectory v=20100 size=618 flags=0x0(none) hashes=22+5 location=embedded
Signature size=4280
Authority=iPhone Distribution: Craig Hockenberry
Authority=Apple Worldwide Developer Relations Certification Authority
Authority=Apple Root CA
Signed Time=Jan 18, 2010 3:28:43 PM
Info.plist entries=19
Sealed Resources rules=3 files=34
Internal requirements count=1 size=144
```

As you can see, the application code is signed with an iPhone Distribution certificate. It's easy to change the signature, too:

```
$ codesign -f -s "iPhone Developer" -vv "Safety Light.app/Safety Light"
Safety Light.app/Safety Light: replacing existing signature
Safety Light.app/Safety Light: signed bundle with Mach-O thin (armv6)
    [com.iconfactory.FlashlightPro]
```

Now, if you display the code signing, you see that it's signed with an iPhone Developer certificate:

```
$ codesign -d -vv "Safety Light.app/Safety Light"
Executable=/Users/studmuffin/Projects/Flashlight Pro/build/Final--Release-
    iphoneos/Safety Light.app/Safety Light
Identifier=com.iconfactory.FlashlightPro
Format=bundle with Mach-O thin (armv6)
CodeDirectory v=20100 size=578 flags=0x0(none) hashes=22+3 location=embedded
Signature size=4290
```

```
Authority=iPhone Developer: Craig Hockenberry (8RMK66WF7S)
Authority=Apple Worldwide Developer Relations Certification Authority
Authority=Apple Root CA
Signed Time=Jan 18, 2010 3:29:16 PM
Info.plist entries=19
Sealed Resources rules=3 files=34
Internal requirements count=1 size=144
```

And that means you can install it on your device with the Xcode Organizer or iTunes. Everything that's in the final build can run on your iPhone, so you can run your final tests while the application is in review.

To learn more about how *codesign* works, make sure to check out its Unix manual page; type *man codesign* at the command line.

Note: In the Flashlight Pro project, resigning the code for the final release is automated with a build phase named Resign Release. This build step runs the *codesign* tool on the application after it's been packaged for release. It also creates a Zip file that contains the modified version of the application.

The First Promotion

It's official: The code for your first release is done. You could sit around twiddling your thumbs while waiting for Apple to approve the application. But you want this product to be a success, so it's time to get started on the next phase of the lifecycle. It's time to start promoting that awesome code!

Some developers use this time to start generating some prelaunch buzz. You could unveil a part of your promotional website with a big "Coming Soon" banner, for example. The risk here is if any delay occurs in the approval process. If your app gets held up for any reason, that buzz you worked so hard to generate will just fizzle out and die. Naturally, people will also ask when "soon" is. You won't be able to tell potential customers or the media when your launch date is because it's completely under Apple's control at this point.

A safer approach is to start a mailing list now. People can sign up to be notified when the release becomes available. Make sure that people know that this is a one-time announcement and that you won't abuse their email address. Everyone gets enough spam as it is. (And if you do use that email address for spam, you'll die in a fire.)

App Review

Once you've uploaded your app with iTunes Connect, you're waiting in line. And it's a line with about 1,000 other developers. The length of the app review process has varied over time, with the average wait time being between a few days and a couple of weeks.

Waiting for your app to be approved is hard: Everything is done and you're anxious to show it off. Time passes more quickly if you focus on other activities like getting your website finished, press releases written, and ramping up your marketing efforts. Now is a great time to contact the media and get them interested in the upcoming release—get as many bloggers and journalists signed up as possible.

The app review process can also be a source of frustration. Don't feel bad if you get rejected; most developers have suffered this fate. Read on to find out what you can do to avoid it.

Make Sure You Haven't Broken Any Rules

There are many causes for rejection. In many cases, it's because you didn't follow the rules laid out in Section 3.3 of the iPhone SDK Agreement that you signed. This section of the document specifies the rules for using the SDK, user interfaces, data, privacy, content, network usage, and other aspects of development. Smart developers read this section carefully.

Unfortunately, much of this legal agreement can be left open to interpretation. The best advice is to use common sense and to do unto Apple as you would have them do unto you.

To help you get a feel for what's acceptable and what's not, here are some well-known sources of rejection:

- **Don't try an end-run around App Store.** You can't offer prizes or any type of monetary promotion in your app. You can't ask for donations from users. Don't use a built-in web browser to facilitate a purchase from within your app. Customers should only buy things from iTunes.

- **Don't mention prices from within your application or on its iTunes page.** As you saw on page 258, prices aren't a simple matter when you take into account that they exist in many currencies. The one exception to this is when you're using in-app purchases: You can query iTunes for a price that can be displayed to the customer.

- **Don't use Apple trademarks or products in your application.** The one that's been most difficult for developers are user interfaces that show a picture of an iPhone. Find a different toolbar graphic or way of showing the device in a help screen. Any other tie-ins with Apple should be avoided: One developer was rejected because of a reference to Steve Jobs; another encountered problems when they used the corporate address, 1 Infinite Loop, as sample data.

- **Don't do anything that makes the iPhone or App Store look bad.** Simulating cracked screens, fake swearing (@%#!?!), and other gimmicks at Apple's expense will lead to disappointment. Apple places a *very* high value on how it's perceived by the public.

- **You need to own all content that's included in your app.** Avoid any unlicensed assets, including mentions of public figures or celebrities. (A royalty-free license to use an asset doesn't mean you own it.)

- **If you have any unfiltered content from the Internet, rate your application 17+.** App reviewers are experts at finding content that's not suitable for children!

- **Don't let Internet connections fail silently.** If you're retrieving data from the network, you need to let the user know if the connection isn't working. App reviewers turn on Airplane Mode when testing your app: It should behave predictably in this situation.

- **Keep your users' privacy in mind.** Make it clear when information from the device, including game scores, is being uploaded to the Web. You don't need to do this with every connection, just give the user a chance to opt out. Providing a setting in your app gives users a chance to change their mind.

- **Don't submit an app that's nothing more than a business card or company advertisement.** These, along with extremely simple joke and novelty apps, will be rejected because of "minimal user functionality."

- **Be careful about duplicating the functionality of Apple's apps.** Developers have had success with applications that replace the built-in utilities, such as the Calculator and Weather apps. The core applications, such as Phone, Mail, and iPod, are generally off limits.

- **Watch the amount of data you transfer on the cellular network.** There's no published limit, but make sure that you don't use more than 5 MB of bandwidth over a short interval (5–10 minutes).

- **Don't include the names of other apps in your keywords.** You're not going to fool anyone if you try to freeload off another developer's success.

- **Don't use undocumented classes or methods.** Yes, it's easy to reverse-engineer these hidden capabilities, but you're creating a compatibility problem with future versions of the iPhone OS. To combat this problem, Apple scans each application after submission to ensure that it's only using published APIs.

These guidelines can change over time: Apple has relaxed some requirements and tightened others. Fortunately, a website tracks the causes of rejections: *http://appreview.tumblr.com/*. It's a good idea to check this site periodically to keep abreast of any changes.

Coping with Rejection

When your application is rejected, you get an email from Apple explaining what's wrong. If there's a problem with the application, you must fix the issue and build a new binary. After testing the new version, use iTunes Connect to submit the updated version.

When a problem arises with your app metadata, such as the description or keywords, you have to edit the information. The next step isn't obvious, but it's essential. You must reject the current binary in Manage Your Applications and resubmit the last binary you uploaded.

Tip: If you find a problem with your application after it has been submitted, use the Reject Binary feature in iTunes Connect. This button halts the review process so you can fix the issue and try again.

After you resubmit the binary, the review process starts again. The most frustrating part about the rejection is that you're at the end of the line again. The wait begins anew.

Ready for Sale

After you click an application in the Manage Your Applications list, you see a link for Status History. Initially, the only items shown are the dates you created the application metadata and uploaded the binary. Eventually, the status will change to In Review, indicating that Apple has started to evaluate your submission. If all goes well, the next item listed is Ready For Sale. Congratulations, your app is ready for worldwide release!

Note: Apple will also send a message to the email address you used for your Apple ID. You don't have to refresh the Status History page obsessively. Or at least not as much.

You have two important tasks at this point. The first is to decide when you want to make the release available. Make sure all your promotional materials are ready to go: the website, press releases, email notices, and anything else you have planned for the launch.

Even if everything is ready immediately, wait a day or two. That seems counterintuitive, but there's one more important thing to do in iTunes Connect—generate promotional codes. This feature becomes available once your app is Ready For Sale. Bloggers and journalists can use these codes to download and install your product before it becomes available to the public at large. It's important to get these individuals interested in your product before you can send them a promotional code. Use the time spent waiting for approval wisely.

To generate the promotional codes, follow these steps:

1. **Log into iTunes Connect and select Request Promotional Codes.**

2. **From the drop-down list, select your product name and then type the number of codes you'd like to generate.**

   The number of remaining codes is displayed under the entry field. Click Continue.

3. **Turn on the checkbox at the bottom of the agreement that's displayed and then click Continue.**

4. **Click the Download Codes link to retrieve the text file containing the promotional codes.**

   Each line of the file contains a 12-character code that can be redeemed on iTunes.

5. **Click Done to return to the main iTunes Connect screen.**

Send these promotional codes out to the media, and give them time to write up their impressions. Each reviewer only needs one code, and they all probably already know how to use it. If not, explain that Redeem links are at the bottom of iTunes' App Store page on the desktop and at the bottom of the Featured tab in the device's App Store app. The codes work just like Apple's gift cards, except they don't cost a dime.

Bloggers and journalists also want to know your availability date so they can schedule their review appropriately. Give them access to a press kit that contains materials they'll need for an article: It should contain a press release, a high-resolution application icon, the screenshots you submitted to Apple, and any other marketing materials you may have.

When dealing with writers, remember that they're probably on a deadline. Make yourself available for any and all questions or problems they might have. Momentum is very important in the early days of a release: Do whatever you can to help these writers to get the word out about your new product.

Launch Day

All the pieces are now in place, and your release date is approaching. As you saw above when you set the Availability Date, the time that your release goes live is sooner than you probably think. Make sure your website is in working order, and wait for the buzz to start.

Keep an eye on your website logs, too. You'll be able to find people talking about your product on Twitter, Facebook, and other social networks. Hopefully, blogs and other online media will start linking to your site. Keep an eye out for sites you haven't heard of, and get in contact with them to say thanks and offer any assistance they might need.

Tip: Now is a good time to check out how your page looks on various iTunes stores worldwide. At the bottom of the main App Store page in iTunes, you'll see a small flag icon. After clicking the flag, you can switch to any store and then navigate to your product page.

It's going to be a long day, but rewarding beyond words.

You've Got Customers!

Y ou've launched your product. Customers are downloading your app from iTunes. Life is good!

But to maintain that good life, you have to work at it. You need to advertise and promote your application to new customers. To understand how effective your marketing effort is, you keep track of sales and other trends. You also begin establishing a conversation with your customer base through social networks and other support channels.

The goal is to gather as much information about your customers as possible. With this data, you learn about their buying habits and what it takes to make them happy. And that makes for a successful business.

Tracking Sales

If you're like most iPhone developers, you want to know how well your product is selling. The more downloads you have, the more people like your work, and you feel validated. More importantly, these sales numbers are the primary feedback mechanism for your marketing efforts. Apple provides two types of reports: a daily summary and monthly financial report.

Daily Reports

The daily summary tells you how many units you sold of each product in each country. It's a tab-delimited text file that totals every item sold. It shouldn't come as a surprise that you get these files from iTunes Connect. To download the report, follow these steps:

1. **Log into iTunes Connect and click Sales/Trends Report.**

   The Transaction Reports screen appears (Figure 9-1).

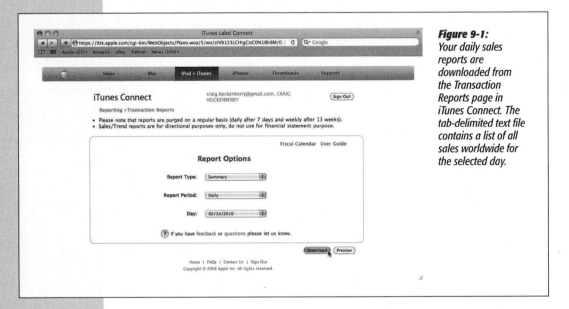

Figure 9-1:
Your daily sales reports are downloaded from the Transaction Reports page in iTunes Connect. The tab-delimited text file contains a list of all sales worldwide for the selected day.

2. **Make sure the Report Type pop-up menu shows Summary. Set the second pop-up menu, Report Period, to Daily.**

   The page updates, and you see some new options, including a Day pop-up menu.

3. **From the Day drop-down menu, choose the date you want reported. Then click Download.**

   A text file for the selected date makes its way to your Downloads folder.

You can load this text file into a spreadsheet such as Excel, and manually track your sales, but it's tedious. After a few days of processing these files, you'll be happy to know that several products on the market make it easy to gather, analyze, and display your sales data. As a bonus, many of these applications can also track your reviews and ranking on iTunes.

Web-based solutions

If you'd like your sales data to be accessible no matter where you are, a web-based solution that shows the information in a browser would be handy. The folks behind appFigures (*http://appfigures.com*) have produced exactly this.

After signing up for an account, enter your iTunes Connect login info, and your daily reports are transferred to the application, analyzed, and plotted. appFigures also provides a very useful "events" mechanism that helps you track the effect of releases and promotions on your product's sales. Reviews and rankings are also available.

You can use the site for free with a limited number of products and accounts. If your needs go beyond this or you wish to use additional features—like reviews and ranking—the monthly fees start at $5.

iPhone apps

If you like to do *everything* on your iPhone (you know who you are), you'll be happy to know about two great solutions for app sales tracking.

The first is called AppSales, and it's open source software written by Ole Zorn. You can check out the project on GitHub: *http://github.com/omz/AppSales-Mobile*. Once you check out the source code on your Mac, you can build and install the application on your iPhone. You then supply your iTunes Connect account information, and the application loads your daily reports and shows the results in tables and graphs. You can also see your products' reviews.

Another solution is the similarly named MyAppSales at *www.drobnik.com/touch/my-app-sales/*. This app provides several features not found in AppSales: It can handle multiple accounts and in-app purchases. Because Apple doesn't let iPhone applications access iTunes Connect per section 3.3.7 of the SDK agreement, you'll also need to build and install this application yourself. The cost for a source code license is a very reasonable $15.

Desktop application

The third solution for doing daily reports is a Mac desktop application called AppViz at *www.ideaswarm.com/products/appviz/* (Figure 9-2). As with the others, you provide your iTunes credentials, and it downloads your reports automatically when you launch the application. This $30 application can also back up your sales reports automatically.

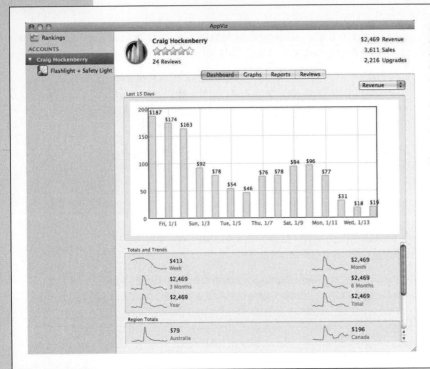

Figure 9-2:
Here's how AppViz shows the daily sales information for Safety Light on the App Store. The AppViz dashboard is plotting the daily revenue for the first 2 weeks of 2010. The application can also produce more detailed graphs and sales reports. And you can check out your product reviews and rankings from all iTunes stores worldwide.

AppViz can also download the rankings for the top 100 apps in each category. If your app is a top seller, the rank helps you see how your product is performing in iTunes. As with other sales-tracking applications, worldwide reviews are available.

The main advantage to using a desktop application is that all your financial data is under your direct control. And as with any financial data, you should make sure that it's backed up and stored in a secure location.

Guesstimation

It's important to keep in mind that the daily reports are reported in each country's currency. A copy of Safety Light earns you $0.70 in the United States and 0.48€ in Europe, so that's the way Apple reports it.

Since most developers are interested in knowing the results in their own currency, the applications listed above adjust the sales figures using the current exchange rates. However, foreign currency markets fluctuate, so the reported values will be off by the time the end-of-month reconciliation is done. The adjusted amounts aren't perfect, but they're fine for gauging how well your products are selling. If you're doing any revenue sharing or paying royalties, don't use the daily reports for the accounting. Instead, use the monthly financial reports, described next.

Monthly Financial Reports

About 2 weeks after the end of the month, you'll start to get emails from Apple with a subject line that begins with "iTunes Financial Report". That's great news—Apple has taken the first step toward paying you.

You'll receive this notification mail for each major region: Australia (AU), Canada (CA), Europe (EU), Japan (JP), United Kingdom (GB), United States (US), and World (WW). It usually takes 4 or 5 days to get the reports for all regions. The email itself doesn't contain a report; you get it by logging into iTunes Connect.

Once in, click the Financial Reports link, and you'll see the reports for the most recent month listed first (Figure 9-3). Click the links in the Download column to get the reports on your Mac.

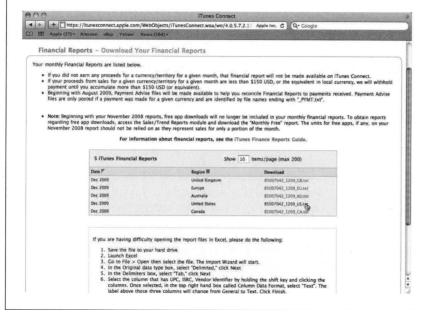

Figure 9-3:
A couple of weeks after the end of the month, you can download Financial Reports from iTunes Connect. When you click a link, you download the report for that region and time period–here, the United States in December 2009.

You receive two types of reports: The files that end in "PYMT.txt" contain a summary of the payment for the region. If you haven't earned at least $150 in a given region, Apple will maintain a balance for your account; any previous balance is listed in the monthly summary. You'll also see any withholding or GST/JCT taxes that are deducted from your earnings.

Note: If you haven't filed Japanese tax treaty documents as discussed on page 237, you'll find that 20 percent of your earnings in Japan are withheld. The Manage Your Contracts section of iTunes Connect explains how to recover these earnings.

The other type of .txt file listed (without "PYMT") on the Financial Reports screen contains the sales information. This document contains tab-delimited data that breaks down the number of units sold for each product in each country. Both Microsoft Excel and Numbers in iWork will open the report after you drag the document onto the Dock icon.

If you have multiple products, sort the data by the Vendor Identifier (to group all products together), and then add up the Extended Partner Share column to get the amount for each product.

Getting Paid

The amounts listed in the PYMT.txt files are what Apple will deposit into your bank account. Typically, the deposit happens a couple of weeks after you've received the monthly report. Reconciling these deposits is a bit tricky because of currency conversions.

Only the US (United States) and WW (World) regions are listed in dollars. The Payment listed in the financial report will show up as an ACH deposit in the account you specified when setting up the contracts in iTunes Connect. For payments that aren't in dollars, you need the exchange rate on the day of the deposit. Apple can't provide this information, because they don't know what the rate will be when the payment is handed over to the banking system, which is when the actual exchange takes place. And unfortunately, many banks don't report the exchange rate for the SWIFT wire transfer.

If your bank doesn't record the exchange rate, use an online archive like the one at *http://x-rates.com/cgi-bin/hlookup.cgi*. The rate for a day or two before the deposit date will help you determine the country of origin and get a feeling for whether the amount is correct.

At this point, all that's left to do is make a down payment on a 100-foot yacht and order the Cuban cigars! Or figure out how to make those monthly deposits a bit bigger. . . .

Advertising and Promotion

Most developers go into iPhone development with expectations that exceed the actual results. You thought you'd make a million bucks with just a few hours of work. Sorry, but it doesn't usually work out that way.

The success of your app depends a lot on advertising and promotion. Customers can't buy your product unless they know about it. Your code is awesome, so don't be afraid to tell people that!

Press Release

The first promotional item to produce is a *press release*. This one page document lets the media know about your new product. As you write your press release, it's important to think about who's going to be reading it. You're writing to the press, not potential customers. The goal is to get coverage for your product that will generate interest that indirectly leads to sales.

So avoid any language that sounds like a sales pitch. Instead, present your product as something that's different and newsworthy. For Safety Light, points that get the press's attention might be:

- It's the best flashlight for the iPhone, unique among the myriad flashlight apps.

- Its rescue beacon can save a life in an accident or natural disaster.

- It has a disco mode.

Another tactic is to use quotes that tell a story about your app: "Safety Light replaced all the other flashlights on my iPhone!" or "My dog just loves the disco lights!" (If these quotes come from an industry expert or minor celebrity, so much the better.) Provide as many talking points as you can to give the journalist a unique angle to write about.

Remember, your press release is not the only one the reporter is going to see. Hundreds of other iPhone apps are being launched on the same day, and many of them are jockeying for the same press attention. It may seem counterintuitive, but a shorter press release is much better than a long, drawn-out screed. Keep it short and punchy, and someone might actually read it.

You should also be aware of the standard format for press releases. The release needs a headline, a summary, the body of information, company background, and contact information. An example, along with other tips and guidelines, can be seen on the PRWeb site: *www.prwebdirect.com/pressreleasetips.php*.

Once you finish your press release, make sure that it's available on your website. Make it easy for people who aren't on your PR mailing list to get this information about your product.

Find a Mouth

Even the best developers tend to shy away from self-promotion. If you're too introverted to learn to talk up your app, you can find someone to do it for you. Either way, your app needs someone out there spreading the word.

Learning how to spread the word about your products can be very rewarding. You'll build new and interesting relationships with smart people throughout the industry. But be warned that this activity eats up *lots* of time. Contacts don't come easily, yet without them you have nowhere to send a press release.

The other option is to find a firm that specializes in press relations or marketing. They already have contacts and can leverage them on behalf of your products. Marketing and PR firms also have a great deal of experience on how to pitch a product. The biggest benefit, however, is that it lets you concentrate on what you do best—software development.

Social Networking: Word-of-Mouth on Steroids

Many developers have found that social networks are an excellent place to increase the visibility of their products. Both Twitter and Facebook let people opt in for regular updates about what you're working on.

Here's where word-of-mouth on steroids comes in. In one status update, you can reach hundreds or thousands of people. To give you an idea of what this looks like, the graph in Figure 9-4 shows Safety Light being announced on a Twitter account with about 6,100 followers.

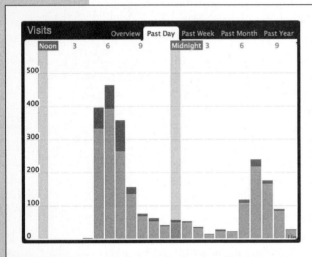

Figure 9-4:
This graph shows the number of visitors to the Safety Light website after its announcement on Twitter. The first peak on December 21 shows the response to the tweet: "Ready for Sale – The best flashlight for the iPhone should be available in iTunes shortly". The second peak, a day later, was from "In case you missed it last night, the best flashlight in the App Store is now available". These two tweets introduced the product to over 2,000 people. The graph was produced by Mint (http://haveamint.com/) running on the Safety Light web server.

Note that returns diminish with repeated announcements. Mentioning a product a few times is fine, but any more than that is going to annoy your followers. Also, consider opening a separate product or business account, so you can keep your personal and business lives separate.

Once you've established your accounts, you want people to follow you. For example, you can put a simple link on your website that visitors can click to follow the Twitter or Facebook account. In fact, it's hard to find a website these days that *doesn't* have social-networking links!

Tip: Your product's web page is also a great place to do cross-product marketing. When someone follows your activities on a social network, it's likely that they enjoy your work. Make sure to introduce them to other products you've created.

The Net works

The best part about social networks is that they can amplify word-of-mouth rec-ommendations. Some of your followers are smarter, more handsome, and funnier than you are. (Hard to imagine, isn't it?) Michael "Rands" Lopp is one such person: He's an author who runs a highly respected website called *Rands in Repose* (*http:// randsinrepose.com*). His @rands account on Twitter has over 11,000 followers.

After seeing one of the previous tweets about Safety Light, he added his own com-ment: "Clearly missing a SafetyDance™ mode: *http://j.mp/7tB9G5*". That link led to over 700 curious individuals learning about Safety Light, as shown in Figure 9-5.

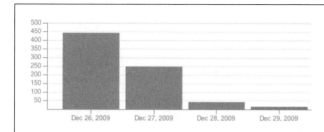

Figure 9-5:
This graph shows the effect of an influential follower mentioning your app. In this case, a tweet from @rands on the day after Christmas led to over 700 people learning about Safety Light. Querying additional information about the shortened link (in the paragraph above) displayed the graph.

Many users on social networks prefer to use shortened URLs (like the link in the pre-vious paragraph), which gives you more room for the content in the status update. In many cases, it also provides you with the ability to gather additional information about the link to your product. You can get statistics about both *bit.ly* and *j.mp* links by adding *info* to the link and typing it in a web browser. For example, entering *http://j.mp/info/7tB9G5* into a browser generated the graph in Figure 9-5. It also provided information on other tweets that used the same URL, sites where the link was clicked, and the geographic location of the person who clicked it.

Give customers something to talk about

Safety Light has a cool feature that's not mentioned anywhere in iTunes or on the product website—the battery graphic that appears when the screen flips to and from settings. The batteries were kept secret on purpose. You can leverage social networks by letting people discover things in your iPhone app. It's human nature to want to tell others about cool things you come across. The first time the user flips over to view settings, he'll be surprised to see the batteries (Figure 9-6). Some will be so delighted that they want to show their followers.

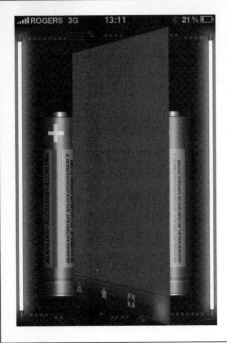

Figure 9-6:
The batteries displayed between the light and settings views are a surprise for new users. Some will comment on social networks and generate additional interest in the app. Martin Dufort (@mdufort) took this screen capture and posted it to his Twitter stream for others to view.

A variation on this theme is when apps post game scores to a player's Facebook and Twitter accounts. Adding this feature to your games isn't difficult, and it lets your users boast about their achievements. If friends and followers haven't heard of the game, they'll check out the included link. If they've already purchased the game, they'll want to beat the score! There's benefit for you as well as for your customer.

Tip: If you want your app to send posts to a social network, be sure that the app asks your users' permission to do so. Every posting should give users a chance to opt out. Automated posts are viewed as spam and break the user's trust in your app.

It's a conversation

For social networks to be truly effective, the communication needs to be a two-way street. If you just pump out what amounts to nothing more than press releases, your followers are going to get bored and leave. The solution: Make sure your social network is interactive.

Here are some tips on how to effectively use your product or company account:

- **Use your account to post tips and tricks about using the app.** Spread out these posts over time so that more users will be likely to see them. It's also a good idea to include links that point to a support page that lists all tips.

- **Have a way to handle support requests directed to your account.** In most cases, you're best off directing the customer to an email account where the problem can be investigated. Complex support issues usually require some back-and-forth conversation to determine the cause and solution, and you don't want all that filling up your wall. Email is best suited for this type of exchange.

- **For short and simple questions, a quick reply does the trick.** For example, a user on Twitter had a question about the light intensity: "@chockenberry Is it possible for the app to override the system brightness? At low brightness, high brightness in your app is still low." A quick response lets them know that you're aware of the problem and that it's a system limitation: "@jazzviolin No public API to adjust brightness, unfortunately."

- **Be on the lookout for new feature requests.** As people use your application, they come up with ideas: Tapping in a short message to your account lets them get the thought to you quickly. A quick acknowledgement lets them know you're paying attention and appreciate the input. Thanks to users on Twitter, a feature request list for the next version includes a BPM indicator for DJs, an avalanche detector using the accelerometer, and custom messages via Morse code.

- **Write a thank-you note.** A great thing about social networks are the compliments: "@chockenberry I love the batteries behind the view when flipping. Nice touch." Thanking your users ("@McCarron Thanks! We try to keep things interesting :-)") not only makes them feel good, but it also makes them even better supporters and champions of your app.

- **Remind followers to leave a review on iTunes.** As you saw on page 259, social proof can do a lot to promote your app; the people interested in your work are usually happy to take a moment to help out. Just be careful how often you make this request: too frequently, and you'll appear to be begging.

- **When you tweet, use easily searchable patterns.** For example, putting "http://safetylightapp.com" in tweets will help you search for people who are talking about the website, while just "Safety Light" will include people talking about bike lights.

The Big Bang

In the last section, you saw how influential people on social networks help promote your application. It gets even better when you're mentioned by a leading website!

This single sentence on December 29, 2009, changed Safety Light's future:

> "If you're going to make an iPhone flashlight app, you might as well make it the best one."
> *http://daringfireball.net/linked/2009/12/29/safety-light*

The actual words aren't as important as the person who's writing them. This short posting occurred on Daring Fireball, an influential website that offers news and commentary about the Mac and iPhone. John Gruber, the journalist who runs the site, has a knack for finding great applications, and readers love his discoveries. As a result of this recommendation, over 10,000 people visited the Safety Light website, and 688 of them decided to purchase the application (Figure 9-7).

The statistics were last updated **Wednesday, 30 December 2009 at 8:48,** at which time **'switch'** had been up for **434 days, 14:09:38.**

`Daily' Graph (5 Minute Average)

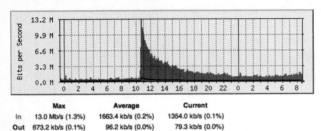

	Max	Average	Current
In	13.0 Mb/s (1.3%)	1663.4 kb/s (0.2%)	1354.0 kb/s (0.1%)
Out	673.2 kb/s (0.1%)	96.2 kb/s (0.0%)	79.3 kb/s (0.0%)

`Weekly' Graph (30 Minute Average)

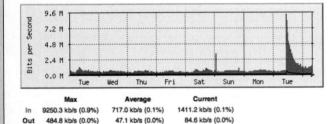

	Max	Average	Current
In	9250.3 kb/s (0.9%)	717.0 kb/s (0.1%)	1411.2 kb/s (0.1%)
Out	484.8 kb/s (0.0%)	47.1 kb/s (0.0%)	84.6 kb/s (0.0%)

Figure 9-7:
This graph shows the network traffic on the web server hosting the Safety Light website. A link from Daring Fireball at 10:30 a.m. caused a huge spike in traffic. When choosing a host for your website, ask about policies for handling bandwidth anomalies like the one shown here. Having a site go down because of popularity will cost you sales.

Prior to this exposure, Safety Light's daily revenue was about $25 per day. On the day of the posting, the daily revenue was over $450. The mention at Daring Fireball also pushed Safety Light into the #40 position of the Top 100 for the Utilities category. The moral: Don't underestimate the importance of your media relationships.

Spreading like wildfire

When a site with a large readership links to a product, the link tends to ripple out to other sites. That's what happened after the mention on Daring Fireball, and it helped extend awareness of Safety Light.

The effect is not unlike the social proof you saw in the last chapter (page 259). If one writer thinks an app is worthwhile, others will take a look and add their unique point of view. That's another reason to have more than one trick up your PR sleeve: One writer will pick up on it being "the best flashlight," another sees the unique design and SOS beacon, while a third latches onto Disco Mode.

Here's a sampling of sites that mentioned or reviewed Safety Light after the Daring Fireball post: A Brazilian site that covers the Mac even picked up the story!

- **Laughing Squid.** *http://laughingsquid.com/safety-light-iphone-flashlight-app-with-sos-signal-feature-for-emergency-rescue/*
- **The iPhone Blog.** *http://www.theiphoneblog.com/2009/12/29/safety-light-iphone-flashlight-app-redeemed/*
- **Download Squad.** *http://downloadsquad.com/2009/12/30/safety-light-not-just-another-iphone-flashlight/*
- **Mac Magazine (Brazil).** *http://macmagazine.uol.com.br/2009/12/30/se-voce-for-baixar-uma-lanterna-para-o-iphone-opte-logo-pela-melhor-safety-light/*

Tip: Google Language Tools (*http://google.com/language_tools*) can help you understand reviews in foreign languages. Parts will be lost in translation, but it's enough to get the gist.

Make sure to get links to reviews onto your promotional site as quickly as possible. Also, pick the best ones and add them to your Application Description in iTunes Connect.

Online Advertising

Another avenue for raising awareness about your app is online advertising. Reviews and other mentions in the press tend to generate buzz for a short period; you'll get a lot of traffic for a few days and then it's over.

Advertising is a great supplement because it's *repeated* exposure. If you advertise on a popular website, visitors will see your brand and marketing message over and over. Also, ad networks let you show ads to users while they're using an iPhone or iPod touch—a direct connection to potential customers.

Budget

So how much do you spend on advertising? To answer that question, you need to look at both the product you're promoting and the type of advertising you want to do.

A general rule of thumb in the advertising world is that 5–10 percent of your gross sales should be reinvested back into product marketing. That's a good starting point for thinking about how much you want to spend, but as you'll see, things change significantly depending on your app's price.

In all these advertising scenarios, think about how much money you're making from each sale. It's easy to forget that Apple's cut is 30 percent. For a 99¢ title, you're earning just 70¢ per unit sold. If your app's price is $4.99, the net revenue is $3.50. As you'll see, finding effective ways to advertise a low-priced app is tough.

Impression-based advertising

This traditional form of online advertising is based on the number of times an advertisement is displayed to visitors. You buy *impressions* in groups of a thousand (commonly called a CPM). Prices for CPM advertising can be as little as $1 per thousand and go up into the hundreds of dollars.

Note: "CPM" stands for "cost per thousand." The *M* in the acronym is the Roman numeral for 1,000. As you enter the world of online advertising, you'll find a lot of strange, three-letter acronyms.

The cost depends on the website that's delivering the ad. Each site attracts different types of customers. The number of visitors they attract each day also varies widely. Don't be afraid to ask for demographic and traffic statistics when inquiring about advertising rates.

If you're going to advertise your iPhone game on a popular gaming website, get ready to pay a high CPM. Smaller sites that serve a niche audience will have a lower CPM and may be a better fit for your product. When looking at costs for impression-based advertising, it's important to keep your break-even point in mind. You'll recover your advertising costs when you divide the cost for the 1,000 impressions by the revenue from each sale.

For example, if you spend $1,000 per CPM of your 99¢ app, you need to sell an additional 1,429 copies just to break even ($1,000 per CPM / $0.70 revenue). Considering that you're reaching a maximum of a thousand customers with the ad run, selling 1,429 copies is highly unlikely. You might as well throw $100 bills out the window. On the other hand, if your app sells for $4.99 and the CPM is $100, you need to sell only 29 copies to pay for the advertising. That means that only 2.9 percent of the thousand people who see the ad need to make a purchase. With the right audience, that's an achievable goal.

As you're looking for sites to advertise your app, remember that the people who are selling the ad space often have flexibility with pricing. Don't be afraid to negotiate the quoted prices. You can often get a better deal with the rate if you're willing to move your ad run to a date that lets the publisher fill an empty slot.

Click-based advertising

Many advertising networks—including AdMob and Google—are now carrying ads based on a cost per click (CPC) model. You bid, in competition with other advertisers, for your ad placement by specifying how much you're willing to pay for each person who clicks the ad. The more you're willing to pay for each click, the more likely and frequently it will show up on the network.

To determine how much you should pay for each click, you need to know how many of the people who visit your page in iTunes actually make a purchase. This *conversion rate* is usually between 5 percent and 10 percent of potential customers. That is, for every 100 people who see your page on iTunes, about 5 to 10 people will buy the product.

To break even when advertising your 99¢ app, multiply your net revenue per unit ($0.70) by the number of units you sell with each click (the 5 percent to 10 percent conversion rate), and you come up with a CPC in the range of $0.035 to $0.07. With a title that's $4.99, you end up with a CPC between $0.17 and $0.35.

Of course, monitor your sales during the ad run to get a better idea of your true conversion rate. The ad network will report the number of times the ad was clicked, and Apple reports on daily sales. Compare the number of sales on the day the ad was run to your daily average, and use the results to compute the conversion rate. For example, an ad for Safety Light had a short run with AdMob on January 5, 2010. Using a CPC of $0.05, 127,634 impressions were generated in just a few hours. The 1,000 clicks on the ad resulted in a CTR (click through rate) of 0.78 percent. Sales for that day were about 20 above average, giving a 2 percent conversion rate. AdMob requires a minimum CPC bid of $0.05, so when a conversion rate on a 99¢ product is less than 7 percent, you're wasting money. Safety Light's conversion rate was well below this threshold, so no more ad campaigns were run.

The fact that the ad campaign was completed in just a matter of hours indicates that plenty of iPhone apps display ads and that plenty of customers are willing to click them. That's good for you, since it means you'll get plenty of exposure even with a low CPC, and that benefits higher priced applications. If Safety Light were priced at $4.99, it would have needed only 15 additional sales per day to be cost-effective with the same 1,000 clicks at $0.05 CPC.

Your goal is to find the lowest possible CPC that generates additional sales, and since each product is unique, the only way to determine that is with trial and error. Spend some of your ad budget to find out what works and what doesn't.

Flat rate

With flat-rate advertising, the website or ad network agrees to run your ad for a certain period at a fixed cost. There's no guarantee for either the number of impressions or the number of clicks. So what does this type of ad delivery have to offer?

Primarily, these types of ads give you high visibility. Your ad is likely to be the only one on the page. One ad network example is the Deck Network (*http:// decknetwork.net*). This network places ads on sites that are very influential in the creative, web, and design industries. If your product appeals to customers in these fields, spending thousands of dollars for an ad run could produce significant results (Figure 9-8).

Figure 9-8:
The only advertisement on this page at Zeldman.com is one for Safety Light. The ad was placed on the Deck Network—a consortium of websites that appeal to the creative, web, and design industries. While the ad was being run, visits to the Safety Light promotional site were about 10 times higher than normal, and sales doubled.

Sponsorship

Another trend that's emerging is *blog sponsorship*. Your product gets high visibility among the blog's readers. As with public broadcasting sponsorship, you're not advertising your product or brand directly. Sponsorship rates range from hundreds to thousands of dollars per week; the more influential the site, the higher the cost. Two examples are Daring Fireball (*http://daringfireball.net/feeds/sponsors/*) and Hivelogic (*http://hivelogic.com/sponsorship*).

The major benefit to these forms of advertising is that your product is associated with a trusted source. As you can see from the response after the mention from Daring Fireball (page 281), it's a lot easier for a customer to buy when your product has this positive association.

Branding

Advertising is expensive. You need to get as much value out of it as possible. Initially, it's easy to focus on advertising a single product, because that's all you have. Smarter developers think about the long term and advertise the product *and* the brand.

You'll see this in action with the larger developers and publishers. If you've been paying attention to the iPhone gaming market, you've probably heard of the popular series Rolando and Topple. What's more important is that you're also probably aware of ngmoco:), the company that released the titles.

That's no accident. Such companies focus on marketing the brand as much as they do the individual apps. The reason is simple: If you're happy with one game from ngmoco:), you're more receptive toward their new releases. You may have no idea what Dr. Awesome is about, but you may buy it just because you loved Rolando so much. Once you have multiple titles, you can build your brand at no cost. That's the best kind of advertising.

Many apps include cross-promotion on a game selection screen or help screen. If a customer is enjoying one app, she'll want to know about your other work. The thing to be careful about when doing this kind of brand advertising is to not annoy your users. Be subtle, or you risk turning a fan into a foe.

When you have multiple titles, also consider a promotional website that showcases all your titles. For example, all the Iconfactory apps are listed at *http://iconfactoryapps.com*. This site extends the cross-promotion to include both Mac and iPhone platforms. The eventual goal is that a customer who sees "from the Iconfactory" thinks "this has got to be good."

Promotion Codes

Another way to get free advertising is with *promotion codes*. As the name implies, these codes can help you increase awareness of your product. As you saw on page 262, reviewers should get the first shot at your promotional codes. By giving a single code to a popular reviewer, you can reach hundreds or thousands of potential customers who read the review.

Tip: Developers, reviewers, and users all tend to refer to these codes as *promo codes.* The shortened version certainly helps when you're posting to Twitter, with its 140-character limit!

Luckily, Apple provides more promo codes than you'll need for the launch—every version of your app gets a total of 50. You probably won't use all of them for reviewers, so use the rest to generate buzz about your product. And the best way to do that is by giving them away. People love to win prizes, even if it's only worth a buck or two.

The simplest way to give away the promo codes is by posting a message to the people who follow your social networking accounts. Advertising is all about reaching people, so how you give away those precious codes is important.

Strings attached

Some developers just post the promo codes and don't ask for anything in return. That certainly works, but you're not going to reach more people than the ones who already follow you. There's no "echo" to pique the interest of others. Instead, make sure that your followers tweet something to their followers in order to win the prize. As an example, this message was sent out to over 6,100 followers on Twitter:

> "Want a FREE copy of Safety Light? Post a new tweet with http://safetylightapp.com URL and I'll DM a promo code to first 20. Be creative!" http://twitter.com/chockenberry/statuses/7491384567

Over the next couple of hours, 60 links to the Safety Light promotional site were posted on the social network. These links, in turn, reached hundreds of people who visited the site (Figure 9-9).

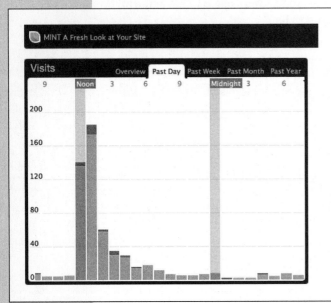

Figure 9-9:
This graph shows the traffic generated by giving away promotional codes on Twitter on January 7, 2010. Traffic to the Safety Light website was higher for about 12 hours, and sales were higher for the next 5 days.

Twitter's search function was then used to look for the winners: *http://search.twitter.com/search?q=safetylightapp.com*. After 20 posts were shown, a direct message was used to send the promo code to the listed accounts. Some developers have resorted to automated means for generating these promotional messages. These *tweet blasts* are a powerful promotional tool, but they carry the risk of alienating potential customers. After all, the same message appears over and over and becomes tiresome. For a more personal touch, ask your followers to be creative when they tweet about your app.

A variation on this giveaway is to offer your promotional codes to a blog as a contest prize. Many blogs like to have these contests to attract new readers and generate excitement among existing ones. You benefit from offering a contest prize because you reach the blog's visitors, who can number in the hundreds or thousands. There's also a positive association with the blog readers, since you're supporting something that they love.

Note: As you start to give out these promo codes, you'll learn about a serious drawback. Unlike every other aspect of iTunes, the codes currently only work in the U.S. store. Some of the winners won't be able to redeem their prize.

Many users outside of the United States already know about this limitation and won't bother entering the contest. For the ones who don't, you'll need to apologize and explain that this is Apple's policy, not yours.

Replenishment

Remember that you get 50 additional promotional codes with each new version that Apple approves. After you see Ready For Sale (page 268), the counter is pushed up to its maximum value, even if you had some codes left on the previous version. Don't let your promo codes go to waste. Request all of your remaining promo codes as soon as you upload your new version. These codes are valid for 30 days, and you can use them for promoting either the current version or the upcoming one.

Sale Prices

Pick up a local newspaper and look at the advertisements: you'll find no shortage of 50 percent OFF or SAVE $20 with LIMITED TIME OFFERS. It's normal for consumers to think of the money they're saving, rather than the money they're spending. Good advertisers know that everyone wants to think they got a great deal with their purchase. The same is true in the App Store. You can increase your sales momentum by giving customers a reduced price.

For many products, you're not going to make a lot more revenue with this maneuver. For example, if you're selling 100 units a day at $1.99, expect to sell about 200 units at 99¢. Your gross revenue won't change much during the sale. So why do it?

- **A sale gives you an opportunity to generate buzz on the Internet.** Sites will link to your product because they know their readers love knowing about good deals.

Note: Remember, those web links will still be around after your sale is over. That's a good thing, because people will still visit your promotional site and learn about the product.

- **Increasing the size of your customer base also helps with word-of-mouth recommendation.** People who love your product *and* got a deal will tell their friends.

- **The increased sales volume can get you into the top sellers for your category on iTunes.** If you're already there, it will raise your ranking.

Sale prices also play a part in determining your launch price. If you start at 99¢, you have no room to play this game. Also remember that it's much easier to drop a price than it is to raise it: Some customers will feel cheated by the increased cost.

Eddie's not really crazy

The ultimate sale, of course, is to give your product away for free. Sounds crazy until you consider the points on page 289. You're essentially advertising at a cost of lost sales. For example, if your normal sales for a 99¢ app are 100 units per day, giving it away for free costs you $70 per day. When you consider the advertising costs presented earlier in this chapter, that's pretty darn cheap.

Note: A variation on this promotional technique, discussed on page 260, is to offer a free trial version that uses in-app purchases to provide additional features. A great trial version will build awareness for your app and lets you sell while customers are enjoying your product.

Let people know

To maximize the effect from sale prices, you need to let people know about it. Yes, that means you're going to be advertising an advertisement. Welcome to the crazy world of marketing!

You'll use your normal press channels to let the media know when the sale is happening. Social networks are also a great mechanism for getting the word out.

For applications that are already in the charts, or that expect to be because of the increased volume, updating the application icon in iTunes Connect can draw attention to the promotion.

Plotting It Out

So how does all of this advertising and promotion affect your revenue? The best way to answer that question is with a graph of a year in the life of Gas Cubby (an app that tracks vehicle maintenance and calculates your mileage). The app's developer, David Barnard, kept close records of his sales and events throughout the year and produced the chart shown in Figure 9-10.

You can easily see the effect of being mentioned on major websites like The Unofficial Apple Weblog (TUAW), Gizmodo, and CNN/Time. Another event that many iPhone developers look forward to is the holiday season: Sales take a big jump as people receive gift cards and shiny new hardware on Christmas Day.

Initially, the App Store showed product updates on each category's page. The spike you see for the 2.0.1 update in Figure 9-10 wouldn't happen today, since only new releases are featured within iTunes. Keeping a product updated may generate some press, but its primary benefit is to keep customers happy by providing bug fixes and new features, not to drive new sales.

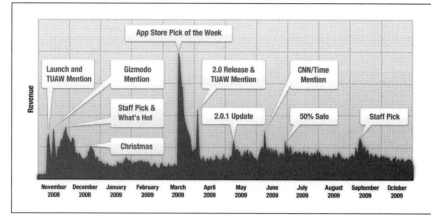

Figure 9-10:
This chart shows the first year of sales for the Gas Cubby app sold by App Cubby. Mentions on prominent sites and new releases produce sales spikes. Promotion within the App Store also has a huge effect on the revenue.

The biggest spikes come from Apple itself. Appearing in the What's Hot or Staff Picks on the front page of the App Store provides a huge increase that lasts over several weeks. The huge spike in the middle is the result of being featured on Apple's website and in a mailing sent out to all customers.

Tip: You can read more about the graph in Figure 9-10 and about many other marketing topics in *Marketing iPhone Apps* by Rana June Sobhany (O'Reilly).

Monitor Coverage

Now that your advertising and promotions are firing on all cylinders, you have to ask yourself: Is it effective?

To answer that question, you have to monitor your product's discoverability, sales ranking, and link effectiveness. That sounds a bit like marketing mumbo-jumbo, but it's actually some pretty cool engineering.

Ranking

Apple provides data feeds that contain the rankings for the top application in each App Store category as well as overall rankings. The feeds are broken down into free, paid, and top-grossing lists. As you learned on page 273, several of the sales tracking applications let you download and display these rankings.

If your product makes it into one of the Top 100 lists, you'll want to track its position very closely. That's when it's time for a Mac desktop application called MajicRank (*http://majicjungle.com/majicrank.html*).

Dave Frampton developed this app to track his popular iPhone game, Chopper. With MajicRank, you can automatically download the ranking at an interval of your choosing. Your application's rank is then plotted on a graph, as shown in Figure 9-11. MajicRank requests a $20 donation for using the software.

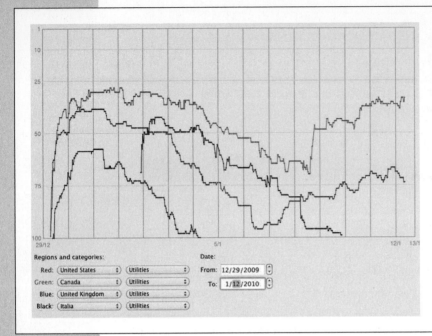

Figure 9-11:
This graph, produced by MajicRank, shows how Safety Light's ranking changed every half hour after being featured on Daring Fireball. You can see the effect of advertising and other promotions that began on January 5. It's interesting to note that the only European chart that Safety Light appears on is Italy and that the only localization for the app is Italian.

Rank tracking is also a good way to keep an eye on your competitors. Unlike the detailed sales figures, which only you and your team can access, rankings are public information. When a competing app does a release with a new feature and its ranking jumps, you can infer that sales have increased. That means you have some work to do.

Google

Google is your friend when it comes to finding links to your app. Of course, you'll want to search for your application name. For a common application name like Safety Light, you'll get a lot of hits for things that aren't iPhone applications. In such cases, just search for *Safety Light app*, and the results will be much more relevant.

Also search for links to your promotional website. Besides typing *safetylightapp.com* into Google's main search box, see which bloggers are linking to your product. To do this, go to *http://blogsearch.google.com* and type *link:safetylightapp.com*.

As you look for people talking about your product, try to think like they do. Use phrases like *flashlight for iPhone* and *flashlight app* instead of just the product name. If you're in the App Store with multiple competing apps, you'll also gain insight into what people are saying about these other products.

Get rich clicks

Wouldn't it be great to know anytime a customer clicked a link that led to your page on iTunes? And wouldn't it be cool to know when that link resulted in a purchase?

Apple has all of this valuable information, and the best part is that you get paid to collect it! You're sending customers to iTunes, so you earn a 5 percent commission for every purchase that results from an inbound link. Just sign up to be an iTunes Affiliate: *http://www.apple.com/itunes/affiliates/*.

Signup

The program is run through a third-party company: You create an account with LinkShare. From this account, you apply to be an iTunes affiliate, create customized links, and view the statistics:

1. **Go to the iTunes Affiliates page (*http://www.apple.com/itunes/affiliates/*), and then click the "Apply to be an affiliate" link.**

    You're transported to LinkShare's signup page.

2. **Select the appropriate Legal Entity Type.**

    If you're a single developer, choose Individual and supply a Social Security Number (SSN). For a businesses, select the appropriate type and supply an Employer ID Number (EIN).

3. **Enter your name and mailing address.**

    LinkShare uses this information to mail you a check for your earnings. You must also supply a unique email address and account information to access the web application.

4. **After clicking Continue, you're prompted to enter the name of your website along with the URL and other information that describes it.**

    If you have a company site, use that, because you won't want to go through this signup process each time you release a new application.

    Since your promotional site is new, you don't know the exact amount of traffic it will generate. Many sites receive between 5,000 and 50,000 views per month, so that would be a good estimate. Also note in the Short Description field that your website is for promoting an iPhone app in the iTunes App Store.

5. **When you click Continue, you should see a confirmation page. At that point, you can click Continue again.**

    You arrive at the Publisher portal.

At this point you've just created an account; you still need to apply to be an iTunes Affiliate:

1. **On the Publisher home page, click the PROGRAMS link in the navigation bar.**

    You arrive on a page where you can search for companies.

2. **In the Advertiser Search box, type *Apple iTunes* and then click Go.**

   You see a single result for iTunes.

3. **Turn on the checkbox on the left, and then click Apply at the bottom of the page. In the pop-up window that appears, accept the terms and conditions, and then click OK.**

   The Status column in the search results will show "Pending" while your application is being processed. You can check the My Advertisers tab to see when the status changes to "Approved".

Both LinkShare and Apple will review your request. In most cases, the application will be accepted in a matter of days. Some developers outside of the United States have been initially rejected for the program: Don't hesitate to contact either Link-Share or Apple by email if you have any problems.

Creating links

As an iTunes Affiliate, you'll have access to a link creation tool. This tool lets you search for an application in iTunes and generate a link with your unique account information:

1. **Go to *http://www.linkshare.com/* and click Login to open the Publisher portal. At the top, click LINKS to display a list of your advertisers; then click Apple iTunes.**

Note: You won't see "Apple iTunes" in the list until your site has been approved. You'll see a page for creating a link specific to iTunes.

---

2. **Near the top of the page, click Link Maker Tool.**

   You arrive at a new window where you can search iTunes for titles.

3. **Since you're interested in iPhone applications, select Applications for the Media Type. Leave the country selection as USA, and enter your application's name as the search term.**

   You're shown a list of matches. To the right of your application's name, you see an arrow link displayed.

4. **Click the arrow to display the HTML for a standard link button.**

5. **It's likely that you want only the address for the link, so just copy the value for the *<a>* tag's *href* attribute.**

The link that's generated is long and hard to read, but you need to understand it before you can take advantage of it. The URL's domain is *click.linksynergy.com,* and the path begins with */fs-bin/stat*. Parameters include the *id,* which is your affiliate ID number, and an *offerid,* which specifies the link is for iTunes. If you look closely, the RD_PARAM1 is an encoded URL for your product on iTunes. That URL also ends with information that ties the link back to your LinkShare account.

But the most important parameter is the one that's not there. Yet.

Monitor activity

LinkShare lets you add one last parameter to the URL that indicates where the link originated. An identifier, referenced as "Member ID" in the Publisher portal and "signature" in other documentation, is represented by the *u1* parameter in the URL. You need to add this parameter manually.

As you choose signatures, make them something that will help you quickly identify the link. For example, if you were running an ad at *CHOCKLOCK.COM,* you'd want to use something like *CHOCKLOCK_AD.*

The link generated by the iTunes Link Maker will end with something like *partnerId%253D30.* To track the links from your ad, you'd append an *&u1=* and the signature to the end of the URL, ending up with *partnerId%253D30&u1= CHOCKLOCK_AD.*

The *u1* parameter should be no more than 72 characters. Stick to using letters, underscores, and numbers: Any other characters might mess up the URL.

Note: These links with embedded signatures are not limited to web pages. You can also use them from within an iPhone app. An article on the iPhone developer site explains several ways to do this: *https:// developer.apple.com/iphone/library/qa/qa2008/qa1629.html*.

As you'd expect, customers will be taken to your app's page in iTunes when they click the link. The total number of clicks for the signature is also recorded in LinkShare's database. Apple will also have the information needed to credit you with a 5 percent commission on any purchase made. Note that this commission is not limited to just your application: If someone buys a song or video within the next 72 hours, you earn a kickback on that sale, too.

LinkShare updates its reports daily. Initially, you see only the Clicks column filled in, because there's a lag as Apple transfers sales information from iTunes back to Link-Share. The commissions can take up to a week to appear in your reports, but when they do, you'll see something like Figure 9-12.

Member ID	Advertiser ID	Advertiser Name	Clicks	Sales	Commissions
SAFETY_LIGHT_BADGE	5318008	Apple iTunes	1,152	1,229.05	61.45
SAFETY_LIGHT_DOWNLOAD	5318008	Apple iTunes	253	202.17	10.11
SAFETY_LIGHT_PRICE	5318008	Apple iTunes	105	98.36	4.92
SAFETY_LIGHT_TW	5318008	Apple iTunes	1	0.00	0.00
SAFETY_LIGHT_TWEET_REVIEW	5318008	Apple iTunes	649	228.53	11.43
Grand Total			2,160	1,758.11	87.91

Figure 9-12:
This table comes from the LinkShare report for links onto the Safety Light page in iTunes from December 29, 2009, to January 5, 2010. It shows that the App Store badge on the promotional site is the most popular link and that asking Twitter followers to leave a review generates sales. The 5 percent commission paid by Apple earned $87.91 for the week.

All of the links on the Safety Light promotional website use the LinkShare signatures. Look at the site and you'll see the names listed in the table at the end of a *very* long URL.

To view the reports, go to the LinkShare Publisher portal and click REPORTS. Then, under Advanced Reports, you have several choices for the Report Type:

- **Signature Activity.** This report summarizes the number of clicks and sales for each Member ID signature that you appended to the link. This overview makes it easy to see which of your links are generating activity and sales. You'll use this report the most often.

- **Signature Orders.** Each transaction in iTunes is listed with the Member ID signature that generated the commission. You'll also see the date and time of the transaction as well as the SKU category that was ordered.

- **Product Success.** To see how much you've earned in each SKU category, select this report. The category "Q9020" shows the number of apps purchased. The number of items can be greater than the number of orders: Multiple applications can be purchased at once.

Tip: LinkShare also provides some of this information with a simple web service. If you'd like to automate the processing of the reports, search for "Signature Orders Report Web Service" in their Help Center: *http://helpcenter.linkshare.com/publisher/questions.php?questionid=780.*

Gathering intelligence

The numbers reported in the Commissions column of the Signature Activity report indicate a nice bit of extra income, but the report becomes much more valuable when you add in the Member ID and Clicks columns. They tell you how much each link is used and how much it earns.

Unfortunately, not enough information is available to know exactly how many of your own products were purchased. Apple and LinkShare don't break down the orders into individual SKUs for the products purchased. If you see one order for $4.99, you have no way of knowing it's for your app and not your competitor's.

Even without this information, you can still determine which links are most effective. Assuming that you have a large enough sample size, you can compare the relative percentages for each signature. The error caused by other products being included will be distributed evenly amongst all signatures. If you see that a signature generates $50 in sales with just three clicks, don't make any decisions based on this partial sample.

Using the table shown above, you can compare the signatures for BADGE, DOWNLOAD, and PRICE. These signatures were used on each of the links on the promotional site: for the "Available on the iPhone App Store" graphical badge, "Download your copy" text link, and "Only 99¢" link in green.

Combined, they totaled 1,510 clicks. The BADGE link's 1,152 clicks generated 76 percent of the traffic from the site to iTunes. The DOWNLOAD and PRICE links produced 17 percent and 7 percent, respectively. Clearly, you want to make sure that the badge is prominently displayed on the site.

Note: The App Store logo must be licensed from Apple. Get details in the App Store Resource Center on the developer site: *http://developer.apple.com/iphone/appstore/marketing.html#storelogo*. You can also download the standard artwork from this page.

Another statistic you can generate from these numbers is the sales per click. For the BADGE, sales of $1,229.05 were generated with 1,152 clicks, or $1.07 per click. The TWEET_REVIEW link only produced $0.35 per click. The reason for that number is that the TWEET_REVIEW link was posted to Twitter reminding followers to review Safety Light. Affiliate links that contain signatures can be shortened using services like bit.ly. If you look at the information for the link *http://bit.ly/info/77QJhk*, you see the SAFETY_LIGHT_TWEET_REVIEW signature at the end of the expanded link. The information page also graphs the link's usage over time.

Obviously, some of the people who followed the link to write a review had already purchased the application and wouldn't be generating any new sales. It's actually surprising that any sales were made at all: After looking at the Signature Orders report, it appears that most of the commissions were generated from other products.

> **Note:** Some signatures get cut off, as with "SAFETY_LIGHT_TW". Email clients, web browsers, copy/paste glitches, and other processing errors can lead to these bad signatures. You'll learn to ignore them.

You can use these techniques in other scenarios, as well:

- **Finding which advertising works best for your product.** Give each publisher a unique signature, and see whether the ad run at TouchArcade or SlideToPlay works better for your product.

- **Test different ad layouts and/or copy.** Is it better to advertise Safety Light with the tag line "The Best Flashlight" or "Save Your Life"? Is it better to feature the application icon or a clever tag line? By running various ads with different signatures, you'll know what works best.

Customer Support

As your customer base grows, you'll find that something else grows: support email. Your customers will have problems, and it's your job to address those issues and keep them happy.

Everything Is Perfekt

You did plenty of beta testing and stomped out a lot of bugs: Everything should be perfect, right? Unfortunately, the answer is no. Something always appears in the 1.0 release. As a smart developer, you should plan on having a 1.0.1 release a week or two after your initial release. As much as you'll probably need it, don't plan a vacation in the weeks following your launch.

This maintenance release will address the major issues that your customers find. You may also want to include some simple feature requests that have been suggested by users.

Crash Course

After your customer installs an application, iTunes will begin to collect crash reports that are created on the device. As explained in Chapter 7 (page 226), these crash reports are generated automatically and contain information that lets you identify the cause. After collecting the reports during sync, iTunes forwards them to a server at Apple.

Each week, your crash reports are analyzed and posted to iTunes Connect for you to download:

1. **Log into iTunes Connect (page 233) and choose Manage Your Applications. Select your application from the list to view its information.**

2. **Click Crash Reports to view the app's reports broken down by the version of the OS being used.**

    If you see "Too few reports have been submitted for a report to be shown", congratulations, your code is working great!

    If you're not so lucky, you'll see the class and method that caused the crash listed in order of how frequently the crash occurs.

3. **Click Download Report to copy the crash report to your Mac.**

    The downloaded files will have a .crash extension.

When you uploaded your final binary to iTunes Connect, you should have saved a copy of your app and the dSYM created during the build (page 262). Make sure Spotlight can locate these files. This Mac OS X search technology lets the tools find and process the files that contain the debugging information.

To view the contents of the crash report, open the Organizer window in Xcode. Then, under IPHONE DEVELOPMENT, select Crash Logs and drag the .crash file you downloaded into the list at the top of the window. After the crash report is processed, you see additional class and method names as well as line numbers that correspond to your source code.

Tip: Use a version control system (VCS), and tag the project each time you do a release. That way, it's easy to retrieve the source code, app, and dSYM files that correspond to the crash that's being reported. The popular Subversion VC tools come preinstalled with the Snow Leopard release of Mac OS X. To learn more, type *man svn* in a Terminal window.

If you're having problems with symbols being resolved, check out the link to Technical Note TN2151 at the bottom of the crash report page in iTunes Connect.

Dealing with Support Email

Once you've worked hard to design, develop, and promote an application that people want to buy, a small percentage of those customers are going to have questions or problems with your app.

With iTunes, many developers find they have a high volume of sales, so even if only a fraction of users ask for help, you can still end up providing a lot of customer support. Develop habits that keep this volume in check.

Some techniques for dealing with support are as follows:

- **Use canned responses for common problems.** Create a document that contains solutions to issues that users frequently encounter. You can then copy and paste from this document to reduce the amount of time spent corresponding with customers.

- **Put your canned responses on your product site,** and update your Support URL in iTunes Connect to point to this FAQ (frequently asked questions) list. Many customers will find the solution to their problem before they need to contact you.

- **Use Twitter or another social network** to let customers know about problems that are affecting many people. Give users an estimate for how long it will take to fix the problem, and tell them when it has been submitted for review. When the fix is released, remind them to update.

- **Start a list of feature requests.** You'll find that a lot of good ideas come from support email and comments on social networks. Keeping that list updated makes it easy when you start on your next version. Some features will be requested frequently, so all you have to say is "Thanks! That's already on our feature request list."

- **Use a single point of contact for support.** It doesn't really matter if you do support on a web-based service, via email, or with some other mechanism. But if you spread your attention across many different channels, you'll go crazy. And so will your customers, because they won't know the best way to contact you. Choose one method of contact and stick with it.

Product Updates

After you've fixed your initial bugs and completed some simple feature requests, it's time to upload your 1.0.1 release. Luckily, this is much easier than the first time, because much of the metadata in iTunes Connect doesn't need to change.

New Info

Obviously, you'll have made plenty of changes in your source code. But one change is more important than all others: You need to update the version number in the Info.plist file.

When you upload a new binary, iTunes Connect checks the value for the version number, and if it's not greater than the current version, the file will be rejected. You'll also want to remember to remove any nonnumeric values from the version number (for example, your beta version number).

It's a good idea to use a consistent pattern for your version numbers. Most developers increment the value according to the breadth and depth of changes made. For example, if your current release is 1.0, you use the following for the next release:

- **1.0.1.** Minor changes in the release. Primarily bug fixes with no new features.

- **1.1.** Used for a release with significant changes including bug fixes and minor new features.

- **2.0.** A major new version with many bug fixes and features for customers.

Welcome Back, iTunes

By now, you're an old pro at iTunes Connect, but there's still a few new things to learn. To update your application, start by updating the metadata:

1. **Log into iTunes Connect and select Manage Your Applications. Then click the icon of the application you want to update.**

2. **Click the Update Application link to start the update process.**

   The first thing you see is a question asking if export compliance has changed. It's likely that the answer is No.

3. **Click Continue.**

   You see a screen where you get to provide updated information about the new version. Compared with your initial submission, you have fewer fields to fill in.

4. **Start by filling in the Version Number field.**

   It should match the version number you assigned in the Info.plist. The Name and Keywords fields are prefilled with the current values; if you want to change them, now's your chance. Once you submit your updated binary, these fields will not be editable.

   The current values for your content rating are also displayed.

5. **Update any values if necessary. Finally, click "Upload application binary later" and then click Save Changes.**

After returning to the application screen, you see that the version number has been updated and the status is "Waiting for Upload". You also see your application icon appear twice on the Manage Your Applications screen.

It's important to note that you shouldn't update your Application Description or any other data at this point. You may have new features to list or updated screenshots, but if you make those changes now, they show for the current version. Instead, wait until the new version is approved before you do these updates.

The last step is to upload a new binary. Do a final build just like you did with your initial release (page 262), and then click Upload Binary. Use the Choose File button to pick the file, and then click Save Changes.

When you're done, you'll see the status as Waiting For Review. Hopefully, it's a short wait!

Upgrades (or Lack Thereof)

If you're a developer coming from another software platform, you probably already know the benefit of software upgrades. In short, you know how many customers you have and how many of them are likely to upgrade. That information, along with the cost of the upgrade, tells you how much money you should spend on development.

Unfortunately, a key part of this equation is missing on the App Store. There's no way to give an existing customer a discount. Each SKU stands by itself. For games and other types of entertainment titles, this isn't a problem. Customers are used to paying for additional game play with the same characters and theme.

Developers of productivity and other more "serious" apps face a difficult decision: Do you give away the upgrade for free, or do you charge for a completely new title? As you make this decision, keep these things in mind:

- You'll have the same data migration problems that you see when a user moves from a free to a paid application. The documents and settings for the current version won't carry over to the new version without a significant amount of effort on your part.

- Users will complain. In one example, Loren Brichter, the author of the popular Tweetie app (*www.atebits.com/tweetie-iphone*), faced quite a bit of backlash when he decided to charge for his 2.0 upgrade. For whatever reason, customers felt a sense of entitlement to a free upgrade, even though the cost was only $3. Put on your flameproof underpants, because people will voice their discontent through social networks and email.

- If you have a relatively small user base and high development costs, charging for an upgrade can make or break your business. If you're selling to a niche market, spend time teaching your customers how ongoing revenue supports the application that they love.

The good news is that if you're considering doing an upgrade, it means you've made some money off your first version. Many products never make it this far, so feel proud about a job well done.

Congratulations!

It seems like just yesterday you were firing up Xcode for the first time to make a simple flashlight application. But as you look back, it's been a long journey, and you've learned much along the way.

If you're like most iPhone developers, as soon as you finish that first app, you're already thinking about the next one. Top researchers have found that the only cure for this disease is prolonged exposure to Xcode. Thank goodness you're now fully prepared to administer that antidote.

So here's to a long and successful run as an iPhone developer in the App Store: You've earned it. Cheers!

Part Four: Appendix

Appendix A: Where to Go from Here

IV

Where to Go from Here

There's always more to learn when it comes to iPhone development. If you find yourself confused about one of the topics covered in this book, check out the following resources. You can turn to books, websites, and discussion forums for answers.

It's also important to remember that iPhone development is a nascent industry. New technologies and techniques are constantly being discovered. Don't be afraid to use the terminology you've picked up in this book in a Google search!

Help with Objective-C

Books

http://developer.apple.com/mac/library/documentation/Cocoa/Conceptual/ ObjectiveC

Many of the books that cover Cocoa also delve into Objective-C. If you're looking for an excellent introduction and complete language reference, *The Objective-C 2.0 Programming Language* published by Apple can't be beat. The glossary and index are especially helpful as you encounter new terminology and concepts. The document is available in PDF format, making it easier for you to print and search.

Web

http://sealiesoftware.com/blog/

If you have *any* question about the internals of Objective-C, Greg Parker has already answered it. When you inevitably find yourself in the bowels of an objc_msgSend crash, you'll end up on this website with a sigh of relief.

Help with Cocoa

Books

http://bignerdranch.com/products.shtml

Aaron Hillegass has been teaching programming for many years at his Big Nerd Ranch. In *Cocoa Programming for Mac OS X,* he approaches the topic with clear examples and an excellent presentation honed from these classes. This book is revered by many longtime Mac developers and, despite its title, is relevant for iPhone developers looking for a deeper understanding of the classes and design patterns shared by both platforms.

http://cocoabook.com/

Cocoa and Objective-C: Up and Running, by Scott Stevenson, takes an in-depth look at both Cocoa and Objective-C. Scott has a long history of providing online tutorials at his Cocoa Dev Central (*http://cocoadevcentral.com/*) and Theocacao (*http://theocacao.com/*) sites. His direct and personal style is easy to read and highly recommended for developers who want to get the most out of the language and framework.

http://www.cocoadesignpatterns.com/

Both Erik Buck and Don Yacktman are longtime Cocoa developers. In *Cocoa Design Patterns,* they examine the object-oriented design patterns used in the frameworks of the iPhone SDK. Through practical examples and sample code, advanced developers will discover the underlying elegance and consistency of the components used to make their iPhone apps.

Web

http://developer.apple.com/cocoa/

This page is Apple's main reference for all things Cocoa. Use it as a launch pad for exploring the framework.

http://cocoawithlove.com

Matt Gallagher's Cocoa with Love site is a never-ending source of useful code and information about the Cocoa framework. Some of his essays delve into the finer points of development for the Mac and iPhone, while others are quite accessible to newcomers.

http://cimgf.com

Written by Marcus Zarra and Matt Long, the *Cocoa Is My Girlfriend* site covers a wide range of topics in both Mac and iPhone programming. It's a particularly good source of information for two of the most important frameworks on the iPhone: Core Animation and Core Data.

http://www.mikeash.com/?page=pyblog/

NSBlog, a site run by veteran Mac developer Mike Ash, covers advanced topics in Cocoa development. It's where the pros go to learn new things.

Discussion

http://www.cocoabuilder.com/

The *CocoaBuilder* site provides a searchable archive of both the Cocoa and Xcode developer's list. The results are displayed in an easy-to-read threaded format. When you're having a problem understanding something in Cocoa, this site is an excellent place to start looking for a solution.

http://lists.apple.com/mailman/listinfo/cocoa-dev

Sign up for the Apple mailing list on Cocoa development. This list is a good place to discuss frameworks, features, and technical issues with Cocoa, not specific issues with the iPhone SDK.

Help with iPhone SDK

Books

http://iphonedevbook.com

Written by Dave Mark and Jeff LaMarche, *Beginning iPhone 3 Development* takes you through the essential parts of the iPhone SDK with step-by-step instructions. A follow-up book, *More iPhone 3 Development,* covers new features in the 3.0 software and more advanced topics.

http://ericasadun.com

Erica Sadun is a never-ending source of sample code, tips, and tricks for iPhone developers. Active since the early days of jailbreaking, she has compiled these recipes into her book *The iPhone Developer's Cookbook.* This book excels at covering new topics quickly with relevant source code.

Web

http://appdevmanual.com

The official site for *iPhone App Development: The Missing Manual.* Additional material that's not covered in the book will be posted here.

http://iphonedevelopment.blogspot.com

The *iPhone Development* site, written by author Jeff LaMarche, is a great place to find tutorials and commentary about the iPhone. Much of the sample code that ends up in his books sees the first light of day on this site.

http://iphoneincubator.com/blog/

Nick Dalton's *iPhone Development Blog* was launched on the first day the iPhone SDK was released. Since then it has become a valuable collection of tips and tricks.

http://iPhoneDeveloperTips.com

As its name implies, the *iPhone Developer Tips* website has hundreds of tips and tutorials that you'll be able to use in your projects. John Muchow, author and long-time mobile developer, updates the site regularly with great information for all developers.

http://furbo.org

This site, written by your humble author, is a place to cover various development topics related to the iPhone. The essay *http://furbo.org/bootstrap* acted as the genesis for this book.

Discussion

http://devforums.apple.com

Apple hosts these discussion forums. In addition to the normal questions and answers for common problems, this site is the only legal venue to discuss SDK releases that are under a Non-Disclosure Agreement (NDA).

http://groups.google.com/group/iphonesdk

This Google group run by Erica Sadun hosts a range of discussions about the iPhone SDK.

Help with Interface Design

Books

http://developer.apple.com/iphone/library/documentation/UserExperience/Conceptual/MobileHIG

Essential guidance when designing your user interface, *iPhone Human Interface Guidelines* is an Apple publication that covers the challenges and best practices for the best experience possible for your customers. You'll want to save the file as a PDF and to keep it as a handy reference as you're prototyping a new project.

http://www.tobyjoe.com/2009/08/my-iphone-user-experience-book/

Toby Joe Boudreax's *Programming the iPhone User Experience* takes the information in the preceding *Human Interface Guidelines* a step further. He covers many of the same issues, but in greater depth, along with code examples and critiques of many popular iPhone applications. Recommended for designers and developers who are looking to get the most out of their user experience.

http://apress.com/book/view/1430223596

iPhone User Interface Design Projects is a survey of 10 popular and award-winning iPhone apps. This book examines the approach taken by each developer and designer to create the final product. As you've seen with Safety Light, the best way to learn is by going through the process.

Help with Xcode

Web

http://developer.apple.com/mac/library/referencelibrary/GettingStarted/GS_Tools/

This "getting started" document on Apple's developer site provides an introduction to Xcode as well as in-depth coverage of the performance tools you can use to improve your iPhone application.

http://developer.apple.com/tools/

This web page contains links for various topics regarding Apple's development tools. If you have questions about the tools, be sure to check out the Xcode-Users mailing list.

http://search.lists.apple.com/

Apple maintains a searchable archive for all mailing lists. You can find information about Xcode, Cocoa, and other technologies in the archives. Make sure to check other options and features using the navigation at the bottom of the page.

http://developer.apple.com/mac/articles/server/subversionwithxcode3.html

This document from Apple explains how to set up Subversion source code management. The instructions include setting up a server and configuring Xcode as a client.

Help with Web Development

Web

http://developer.apple.com/safari/library/referencelibrary/GettingStarted/GS_iPhoneWebApp

This page on Apple's developer site is a great launch point for both beginning and advanced web developers who want to explore the technical and design issues with Safari on the iPhone.

http://www.alistapart.com/articles/putyourcontentinmypocket/
http://www.alistapart.com/articles/putyourcontentinmypocketpart2/

This series on the popular *A List Apart* web magazine covers the basics of developing for the Safari web browser on the iPhone.

http://jqtouch.com

Advanced web developers will certainly want to check out David Kaneda's plug-in for jQuery. This toolkit lets you design sophisticated user interfaces with nothing more than standard HTML, CSS, and JavaScript.

Discussion

http://groups.google.com/group/iphonewebdev

This Google group is a great place to connect with other developers working on iPhone-specific projects.

Keeping Up with News and Business

Books

iPhone Marketing (O'Reilly)

Rana June Sobhany gives an in-depth view of what it takes to market your app. Through scores of interviews with successful developers, you'll learn what it takes to get your product noticed.

Web

http://daringfireball.net

John Gruber's popular website, *Daring Fireball*, takes a pragmatic approach to tech journalism. Insightful analysis and hype-free reporting make it easy to stay up to date with the latest Apple news and to learn how it affects your business.

http://www.mobileorchard.com/

Mobile Orchard, run by Dan Grigsby and Ari Braginsky, offers news, tutorials, and podcasts featuring interviews with fellow iPhone developers. Their weekly news roundup is a great way to stay on top of what's happening in the industry.

http://blogs.oreilly.com/iphone/

The *Inside iPhone* site is O'Reilly Media's hub for all things related to the iPhone. You'll find an active discussion forum along with frequent blog updates covering a variety of topics.

Discussion

http://groups.google.com/group/iphonesb

iPhone Software Business is a Google group for "small, independent iPhone developers who want to talk to other developers about the business of iPhone app development." You'll find a variety of discussions covering marketing, customer support, and independent consulting.

Open Source Resources

A thriving community of developers works on the iPhone. As a result, there's a lot of excellent open source. Besides being a great way to save time implementing your project, novice developers can learn from the talented and generous individuals that created it.

http://joehewitt.com/post/the-three20-project/

Three20 is an extensive open source project by Joe Hewitt that includes many of the classes used in Facebook applications for the iPhone. You'll find a photo viewer, controllers, and views for composing messages, along with various other utilities such as an HTTP disk cache. The project includes a sample application that shows how to use each component. If you're just starting out, you can learn a lot just by reading the code of this expert developer.

http://touchcode.com/

TouchCode is an award-winning project started by Jonathan Wight. An active community of developers has contributed code such as XML and JSON parsers, an embeddable HTTP server, user interface views, and much more. This code is being used in many iPhone apps, so save yourself some time and check it out!

http://code.google.com/p/kissxml

Unlike TouchCode, which offers a lightweight XML parser, *KissXML* provides a rich parser and the ability to generate XML programmatically.

http://code.google.com/p/core-plot/

If your application has the need to plot graphs, you'd be wise to check out the *Core-Plot* project. This framework integrates tightly with other frameworks such as Core Animation and Core Data to give you a flexible system for turning your numeric data into compelling 2-D visuals.

http://github.com/devinross/tapkulibrary

The *Tapku Library,* written by Devin Ross, is a beautiful collection of user interface classes. You'll find code to do CoverFlow, calendars, graphs, loading views, and much more. His weblog at *http://devinsheaven.com/* is also a great source of information for iPhone developers.

http://cocos2d.org/

The *cocos2d* framework allows you to create rich graphical scenes that include sprites. Perfect for 2-D games, demonstrations, and other types of applications that require coordination between images and user interaction.

http://allseeing-i.com/ASIHTTPRequest/

The *ASIHTTPRequest* project provides an easy-to-use wrapper around the Core Foundation network API. If your project interacts with a web server, submits multi-part form data using POST, or communicates with REST-based APIs, this code can simplify your work. Plenty of good sample code and documentation gets you up to speed, too.

http://github.com/AlanQuatermain/iPhoneContacts

Many developers find the C-based AddressBook framework tedious to work with. Developer Jim Dovey has written several Objective-C classes that make working with your user's contact information much simpler and easier. You may also want to check out Jim's toolkit of useful utility classes at *http://github.com/AlanQuatermain/AQToolkit*.

http://github.com/ldandersen/scifihifi-iphone

Buzz Anderson tackles another difficult area in the iPhone SDK: the keychain used to store secure information. The APIs available in the Simulator and devices differ slightly, so Buzz's *SFHFKeychainUtils* wraps it all up in a class that can be used on both platforms. Also check out his useful user interface elements.

Index

Symbols

Get even more for your money.

Join the O'Reilly Community, and register the O'Reilly books you own.It's free, and you'll get:

- 40% upgrade offer on O'Reilly books
- Membership discounts on books and events
- Free lifetime updates to electronic formats of books
- Multiple ebook formats, DRM FREE
- Participation in the O'Reilly community
- Newsletters
- Account management
- 100% Satisfaction Guarantee

Signing up is easy:

1. **Go to: oreilly.com/go/register**
2. **Create an O'Reilly login.**
3. **Provide your address.**
4. **Register your books.**

Note: English-language books only

To order books online:

oreilly.com/order_new

For questions about products or an order:

orders@oreilly.com

To sign up to get topic-specific email announcements and/or news about upcoming books, conferences, special offers, and new technologies:

elists@oreilly.com

For technical questions about book content:

booktech@oreilly.com

To submit new book proposals to our editors:

proposals@oreilly.com

Many O'Reilly books are available in PDF and several ebook formats. For more information:

oreilly.com/ebooks

O'REILLY®

Spreading the knowledge of innovators   www.oreilly.com

Buy this book and get access to the online edition for 45 days—for free!

"The Missing Manual series is simply the most intelligent and usable series of guidebooks..."
—KEVIN KELLY, CO-FOUNDER OF WIRED

iPhone App Development
the missing manual
The book that should have been in the box

O'REILLY®   Craig Hockenberry

iPhone App Development:
The Missing Manual
By Craig Hockenberry
April 2010, $39.99
ISBN 9780596809775

With Safari Books Online, you can:

Access the contents of thousands of technology and business books

- Quickly search over 7000 books and certification guides
- Download whole books or chapters in PDF format, at no extra cost, to print or read on the go
- Copy and paste code
- Save up to 35% on O'Reilly print books
- **New!** Access mobile-friendly books directly from cell phones and mobile devices

Stay up-to-date on emerging topics before the books are published

- Get on-demand access to evolving manuscripts.
- Interact directly with authors of upcoming books

Explore thousands of hours of video on technology and design topics

- Learn from expert video tutorials
- Watch and replay recorded conference sessions

To try out Safari and the online edition of this book FREE for 45 days,
go to *www.oreilly.com/go/safarienabled* and enter the coupon code BSBNJFH.
To see the complete Safari Library, visit safari.oreilly.com.

O'REILLY®

Spreading the knowledge of innovators   safari.oreilly.com